A newscientist GUIDE

DARWIN UP TO DATE

Edited by Jeremy Cherfas

Design	Chris Jones/Colin Brewster
Production	Jackie Wilson
Cover	Lawrence Mynott
Illustrations	Neil Hyslop
Series Editor	Colin Tudge

Published by IPC Magazines Ltd
King's Reach Tower, Stamford St
London SE1 9LS

Printed by Arthurs Press Ltd
Woodchester, Stroud
Glos GL5 5PB

To order further copies and for information about other *New Scientist* titles, write to
The Publisher, New Scientist
New Science Publications
Commonwealth House
1-19 New Oxford St
London WC1A 1NG

ISBN 85037 549 5

CONTENTS

FOREWORD

CHARLES DARWIN provided biology with a very precious gift. He offered an explanation both for the vast diversity of living things that inhabit the Earth and for the inherent similarities among all that diversity. Natural selection, the preservation of those types best able to prosper in a particular set of circumstances, is the strong thread that unifies all biology. Competition, the inexorable struggle to survive, will see to it that anything that can gain a temporary respite from the struggle by exploiting a hitherto empty niche will prosper, and out of this comes the diversity of life. Descent, the moulding of a new type from variations on an old, will result in similarities that make sense of the diversity.

Darwin did not invent evolution. The facts of the matter were there for all to see, and many people did not accept the biblical fixity of species even before Darwin's time. What he invented was a theory that explained evolution. He replaced the benign intervention of a wise Designer with the blind operation of two supreme forces; variation and selection. In so doing, he changed the way people perceived their world.

Darwin knew nothing about the machinery of inheritance, for Gregor Mendel's work had not been done when Darwin wrote *The Origin* and remained unknown until the beginning of the 20th century, and so his theory lacked a convincing mechanism. With the rediscovery of Mendel and the blossoming of genetics, natural selection gained the mechanism that had been missing. The Modern Synthesis brought genetics and population biology together to create neodarwinism, which provided natural selection with the means to bring about the ends that Darwin himself documented. The science seemed complete with the advent of neodarwinism.

There remain, however, disagreements within biology about the details of Darwin's theory. These disagreements, evidence of a healthy science striving for a more complete understanding, have been seized upon by those who would deny all of Darwin and modern science. This book brings together articles culled from the pages of *New Scientist* to document the scientific debate. One thing is clear. Today, a hundred years after his death, the ideas of Charles Darwin remain the mortar that bind the bricks of biology.

Jeremy Cherfas

PART I
INTRODUCTION

In this section we answer three fundamental but different questions. Just what is the evidence for evolution? Is there any substance to the creationist attacks on Darwin? What can evolution and natural selection tell us about the origin and ancestry of our own species?

Chapter One

Who doubts evolution?

Palaeontologists disagree about the speed and pattern of evolution. But they do not doubt that evolution is a fact. The evidence for evolution simply does not depend upon the fossil record

MARK RIDLEY

Bruce Coleman

Dolphins (top) are mammals. Sharks (bottom) are fish. But both swim fast, and so have evolved the same shape

SOME palaeontologists maintain that animals have evolved gradually, through an infinity of intermediate stages from one form to another. Others point out that the fossil record offers no firm evidence for such gradual change. What really happened, they suggest, is that any one animal species in the past survived more or less unchanged for a time, and then either died out or evolved rapidly into a new descendant form (or forms). Thus, instead of gradual change, they posit the idea of "punctuated equilibrium". The argument is about the actual historical pattern of evolution; but outsiders, seeing a controversy unfolding, have imagined that it is about the truth of evolution—whether evolution occurred at all. This is a terrible mistake; and it springs, I believe, from the false idea that the fossil record provides an important part of the evidence that evolution took place. In fact, evolution is proven by a totally separate set of arguments—and the present debate within palaeontology does not impinge at all on the evidence that supports evolution.

In 1972 two American palaeontologists, Niles Eldredge and Stephen Jay Gould, claimed that the fossil record of any one species of animal is usually as follows. An ancestral form appears in the sediments, as a fossil, fully formed. Fossils of the same species collected from later deposits show no evolutionary change; but then the fossil species simply disappears. It may be replaced by another type, in later sediments, which the palaeontologist may infer is a descendant; but he cannot generally observe any gradual evolution from one to the other. Eldredge and Gould suggested that during the time in which new species arise—the time of "speciation"—evolution is exceptionally rapid; and that these rapid bursts of change ("punctuations") interrupt long periods of relative stasis ("equilibrium"). The alternative, and more traditional view, is that evolution is no more rapid during speciation than at any other time. There is a controversy between these views (see chapters seven and eight).

The twist came when Eldredge and Gould's "facts" about the fossil record were picked up by the creationists and others, and used to fuel the wholly different controversy about whether evolution has ever occurred at all. The two controversies have been badly confused in recent popular presentations. The palaeontological controversy has been portrayed, in television programmes as well as in several journalistic articles, as part of a general Darwinian malaise. We read of "the new clues that challenge Darwin", of "the trouble with Darwin", or "theory of evolution under fire", and we are asked "Did Darwin get it wrong?".

Someone is getting it wrong, and it is not Darwin; it is the creationists and the media. But why?

One reason that keeps betraying itself is that a lot of people just do not know what evidence the theory of evolution stands upon. They think that the main evidence is the gradual descent of one species from another in the fossil record. Thus, on the BBC television *Horizon* programme "Did Darwin get it wrong?" in March 1981, a creationist told us that the facts about the fossils are contrary to the predictions of the evolutionists—but agree with those of the creationists. He was followed by another critic, who told us that the only satisfactory evidence for evolution would be the discovery of a complete phylogeny (that is, a continuous series of ancestral and descendant species). It looked as if, with the announcement—or rather, forcible re-announcement—that the fossil record shows unchanging species separated by gaps, the evidence for evolution had crumbled away.

However, the gradual change of fossil species has *never* been part of the evidence for evolution. In the chapters on the fossil record in *The Origin of Species* Darwin showed that the record was useless for testing between evolution and special creation because it has great gaps in it. The same argument still applies. Eldredge and Gould pointed out the fossil record might be even less complete than Darwin had thought. Populations in the process of speciating are probably small and geographically separated from their ancestral population, so the full course of speciation would not be preserved at any one site of fossil deposition. What we would see is a species replaced by another, obviously related and yet with no gradual intermediate forms. In any case, no real evolutionist, whether gradualist or punctuationist, uses the fossil record as evidence in favour of the theory of evolution as opposed to

Syndication International

Great dane and miniature poodle, vastly different members of the same species

Primula kewensis *created by hybridising* P. verticillata *(below left) and* P. floribunda *(below right)*

J. K. Burras/Oxford Scientific Films

special creation. This does not mean that the theory of evolution is unproven.

So just what is the evidence that species have evolved? There have traditionally been three kinds of evidence, and it is these, not the "fossil evidence", that the critics should be thinking about. The three arguments are from the observed evolution of species, from biogeography, and from the hierarchical structure of taxonomy. Let us examine the claims and details of each and see how they argue for evolution.

Charles Darwin decided to call his mechanism natural selection by analogy to the artificial selection of the animal breeder, and artificial selection is the method of the experimental Darwinian. As an example, let us consider the simple case of artificial selection for high body weight in mice. The experimenter weighs the offspring and permits only the heavier mice to breed. In the next generation, if body weight is heritable, the average weight of the mice will be higher. Thousands of characteristics in dozens of species have been successfully changed by artificial selection. When artificial selection has been carried out for long enough, it has produced changes much greater than the minimum difference between species. A bulldog and a poodle, or a St Bernard and a pekinese, are physically much more different than a wolf and a jackal (which are different species) or even a wolf and a hunting dog (which belong to different subfamilies). There was no creative input to the dog species while it was evolving under artificial selection; the genetic variation already present in the population was all that was needed.

Artificial selection is experimental evolution; natural evolution has been studied as well. The best known examples have been provided by the school of ecological genetics, which studies natural changes of gene frequencies in natural populations. H. B. D. Kettlewell's work on the change of the peppered moth ***Biston betularia*** from peppered white to brown coloration during the past century is the most famous of many such studies (see chapter six).

"This is all very well," I can almost hear a critic saying, "but it's just tinkering. No one has ever seen the formation of a new species." The answer to this depends on the critic's concept of species, but either way he is wrong. If we take first the "morphological" species concept—that species differ by virtue of their anatomical characteristics—then we have already seen that he is wrong. The changes produced by artificial selection, for instance in the evolution of domestic dogs, are easily large enough for the different forms to count as different morphological species. However, our critic is more likely to accept the "reproductive" species concept—that members of a species are able to interbreed successfully and produce fertile offspring—in which case he is even more wrong. Hundreds of new plant species have been experimentally manufactured. I will use the most famous example, the flowerpot primrose *Primula kewensis,* to explain the method.

Two species, *P. verticillata* and *P. floribunda* are hybridised. The hybrid offspring, as expected, are sterile. However, if the hybrids double their number of chromosomes (and this can be encouraged by the chemical colchicine) they are fertile among themselves though not with either of the parent species. *P. kewensis,* in other words, is a new reproductive species. Many strains of common garden flowers, such as tulips, crocuses and irises, are artificial hybrids. Furthermore, this method of speciation is no mere bizarre horticultural or experimental curiosity but has been a major mode of speciation in natural plants. In many genera of flowers, the different species have different simple multiples (*2n, 4n,* etc) of a "basic" (*n*) number of chromosomes. If, as seems likely, these species were formed by interspecific hybridisation followed by chromosomal multiplication, then about a third to a half of plant species must have originated in this way.

The obvious way to test whether a natural species originated by the hybridisation of a particular pair of species is to try to re-synthesise it experimentally. This has now been successfully done for several species. The mint *Galeopsis tetrahit* was the first: it can be artificially made by crossing *G. pubescens* and *G. speciosa.*

We will leave our critic, walking among the many-coloured hybrids in a horticultural garden, his senses being pleased while his mind, we hope, is disabused of one of the commoner fallacies of the evolutionary controversy, that no one has ever made a new species.

If you study a species in detail across its geographic range you find that it varies slightly from place to place. The spatial rate of change is small, but the extreme forms are often so different that if taken alone they would be classified as subspecies, or even as different species. Usually the extremes do not meet in nature so we do not know whether they would interbreed. In some exceptional cases the extremes do meet, forming what is known as a ring species. The herring gull and the black-backed gull co-exist in Britain, but do not interbreed. They are connected, however, by a more or less continuous series of interbreeding populations distributed around the North Pole. If you move west from Britain, across North America, the gulls look more like herring gulls. They continue to vary as you make your way around the North Pole until by the time you reach Siberia they are starting to look more like black-backed gulls. When the extremes meet, as they do in Britain and Northern Europe, they do not interbreed. There is a kind of unilateral speciation.

The evidence of biogeography and artificial evolution alone make a strong argument for evolution. The creationist can hardly suppose that all the infinitesimal geographical variants were independently created. If that argument were pursued he would soon have to defend the view that all individuals were independently created, for all individuals are genetically unique. The only known mode, however, whereby new individuals are formed is by reproduction, not separate creation. In the cases of artificial selection and ecological genetics, we *know* that the change was not produced by the creation of new variants. There is a slippery slope from individual variation, through geographical variation, through speciation, to the higher categories. At some point on this continuum (usually at the species stage) the creationist has to lower his barrier, saying "up to here evolution can proceed, beyond there must be creation". Wherever it is lowered, the barrier is absurd. There is nothing exceptional about the species stage of evolution, and it is not clearcut at all. It varies geographically and can be altered experimentally.

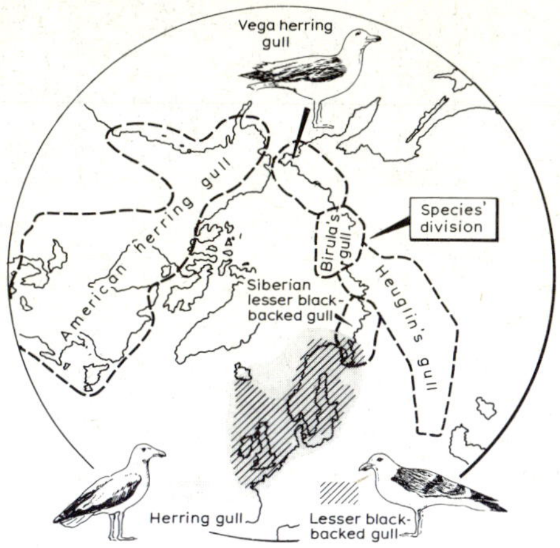

A ring species of gulls surrounds the Pole. The change between neighbours is gradual, but in Europe the two ends of the ring are distinct species

In order to make a comprehensive theory, the evolutionist also needs the principle of *uniformitarianism.* This is the idea that if we can see a process working for a short period of time or region of space to produce a small change, then if we watched it for longer we would see more extensive changes of the same kind. Uniformitarianism is not an empirical principle; it is trusted because of its obvious logic. It is a principle that is necessary to make any historical science possible. So it is not only the theory of evolution that stands or falls with uniformitarianism, but other sciences like geology and cosmology as well. It is not very different from the logic whereby all sciences extrapolate experimental results to nature.

The simple fact that species can be classified hierarchically into genera, families, orders, classes and so on, is not in itself an argument for evolution. It is possible to classify any set of objects into a hierarchy whether their variation is evolutionary or not. It is the *kind* of characters on which the biological taxonomy is based which implies that the hierarchy is the result of common ancestry rather than the clustering of independently created species. The taxonomic characters shared by species are not all shared because of common needs for survival. The point is most easily made by an example. It is a rather exceptional one to use, but it makes the point very clearly. My example is the universality of the genetic code. DNA carries its information in the form of a four-letter code; combinations of these letters stand for different amino acids. The genetic code is arbitrary in the same way that the English language is.

There is nothing about a lion which demands that it is called a lion rather than any other sequence of letters. There is nothing about a glycine molecule which causes it to be encoded by the base triplet GGC rather than any other. However, although all humans do not share the same language, all organisms do share the same genetic code. (The few minor differences in the genetic code of mitochondria do not affect my argument.) If all species had been created separately there is no reason why they should all have been created with the same genetic code. The more likely explanation is that all organisms derive their code by descent with modification from a common ancestor. The example is important because it suggests that all life on Earth had a single origin.

These three are the clearest arguments for the mutability of species. Other defences of the theory of evolution could be made, not the least of which is the absence of a coherent alternative. Darwin's theory is also uniquely able to account for both the presence and absence of design in nature. Animals are generally well adapted but may also have vestigial organs that are no longer functional.

If the creationists want to impress the Darwinian establishment, it will be no use prating on about what the fossils say. No good Darwinian's belief in evolution stands on the fossil evidence for gradual evolution, so nor will his belief fall by it. What we want to know instead is what the creationists say about organic variation in space and time, in nature and the laboratory, about the artificial manufacture of new species, about ring species of gulls, about the universality of the genetic code. And about the host of other cases of which these are but examples.

Chapter Two

Creation science: the ultimate fraud

As an alternative to evolution, creationism is intellectually bankrupt

MICHAEL RUSE

CHARLES DARWIN was an undergraduate at Cambridge from 1828 to 1831. While there, he made friends with some of Britain's most influential scientists: John Henslow, the botanist, William Whewell, physicist and general man of science, and most particularly, Adam Sedgwick, the best field geologist of his day. All three of these men were more than mere scientists. They were Anglican clergymen, and deeply committed Christians. Expectedly, therefore, there were strong connections between their science and their faith. For instance, they believed that all living creatures were first put here on Earth through the direct miraculous intervention of an all-powerful God.

But although these scientists mixed their religion and their science, at the time that Darwin went to university—a full 30 years before he was to publish *The Origin of Species*—they (and other active scientists) had gone beyond a crude belief in the literal truth of the *Bible.* Sedgwick openly confessed that he had given up his earlier support of a universal flood. The geological record speaks against it. And he and the others were starting to think in terms of a very old Earth. How old, no one dared say; but it certainly went back way before the traditional starting date of 4004 BC—a figure calculated by the 17th-century Archbishop Ussher from the genealogies given in the *Bible.*

Given this pre-Victorian renunciation of strict Biblical literalism, *Genesis* was obviously in no position, after the appearance of *The Origin* in 1859, to mount a major threat to the advance of science in Britain. Philip Gosse, clinging desperately to his Plymouth Brethren literalist beliefs in suggesting that fossils were put in the rocks by God to test our faith, was laughed at by all. Most importantly, no concerted attempts were made to teach *Genesis* as science in the state schools. Indeed, some of those most fervently involved in the development of state education—T. H. Huxley for instance—were also those most ardently pushing evolutionism.

In the Deep South of the United States, matters were otherwise. People look to simplistic, comforting doctrines in times of stress, and the period following the Civil War certainly qualified as such a time for Southerners. The plain words of the *Old Testament,* both the factual claims of *Genesis* and the moral claims of *Leviticus,* appealed tremendously. Thus there developed a distinctively American literalist movement: a movement better known today as "fundamentalism".

This movement had its greatest success in the 1920s when several states, including Tennessee, enacted laws prohibiting the teaching of evolutionism in schools. However, as is well known, the fundamentalist victories backfired somewhat. A young teacher, John Thomas Scopes, let himself be charged for teaching evolution. With Scopes being prosecuted by William Jennings Bryan (three times a presidential candidate) and defended by the noted free-thinker and brilliant advocate Clarence Darrow, matters took on a carnival air when Darrow cross-examined Bryan on the literal truth of *Genesis.* Were the six days of Creation really exactly 24 hours each? Bryan could give no convincing reply. Although Scopes was found guilty, evolutionists won a major moral victory. Even more pertinently, the nation laughed at Tennessee, thanks to the savage reporting of H. L. Mencken, who referred to its good citizens as "anthropoid rabble". Such ridicule made other states far more wary of "monkey laws", and so the fundamentalist anti-evolution campaign ground to a halt.

Paradoxically, it was the Soviets who were responsible for its revival. Twenty-five years ago, America was shocked and frightened by Sputnik. Its very existence was taken to be clear evidence of the failure of capitalism and democracy. American science and technology were second-rate. Accordingly, huge sums of money were thrown at the problem, and this resulted in a revamping of American school science education. One cannot teach without guides and aids, so new up-to-date textbooks were commissioned and written.

Naturally enough, the biology textbooks took evolution for granted. But, given the government support of such books and their consequent success, fundamentalist flames were rekindled. And unfortunately, there proved to be plenty of dry wood on which such flames could feed. In the past decade, when we are again living in a time of perceived uncertainty—a function of such diverse phenomena as the breakdown of traditional family units and the rise in oil prices—the new literalist movement has succeeded beyond anyone's hopes or fears.

In some respects, we have moved on today from the 1920s. In 1967, the US Supreme Court finally overturned the Tennessee law, stating that it is indeed constitutional to teach evolution. Additionally, over the years, the Court has ruled that the First Amendment separation of Church and State means precisely that. One may not teach religion *as religion* in state schools. Specifically, one may not teach *Genesis* in school biology classes.

Fundamentalists have therefore been forced into a new tack. No longer is the cry to keep evolution out. Rather, the cry is to get *Genesis* in. But because *Genesis* cannot come in as religion, it must come in some other way. How can this be done? The fundamentalists think that they have the answer. They argue that all of the claims of *Genesis,* taken literally, are complemented by identical conclusions derived according to the best principles and methodology of science. In short, "creation science" (or "scientific creationism"), a body of claims that mirrors *Genesis* entirely, may legitimately be taught in the science classrooms of the nation.

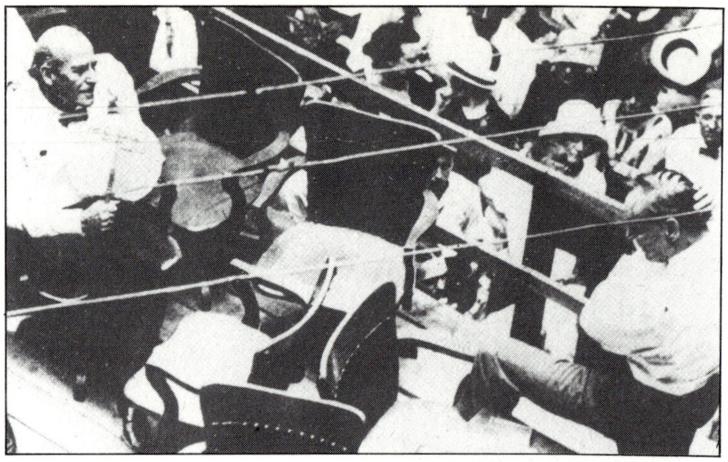

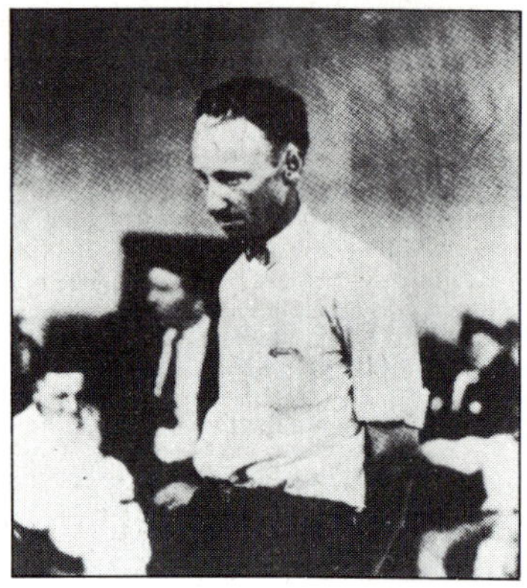

The Scopes trial: Darrow (above right) examines Bryan in front of 2000 people. Scopes (right) was found guilty of teaching evolution

Thus, the demand today is for "equal time" or "balanced treatment" for evolution and creation science in biology classes. This is a demand endorsed by many, including the incumbent president of the United States, Ronald Reagan. Indeed, so powerful is the appeal of the cry for equal time that some states of the Union have already passed laws giving the request legal backing. Arkansas was the first state in line, and thus was the first to have the constitutionality of its laws challenged in court. In fact, the Federal Judge trying the case ruled (in January 1982) that the law is unconstitutional. Nevertheless, similar laws in other states are still on the books, and there are promises of yet more. Law offers one tactic; creationists also intend to press their case less formally, by influencing schoolboards and by approaching teachers.

What is it about creation science that makes an evolutionist like myself think it a dangerous sham? It is not its religious origins *per se.* Rather, it is the dishonest way in which religion is being fobbed off on us as science, and the wretched arguments used to achieve this end.

Take the most striking feature of the whole body of creation science literature, namely the distinctively peculiar fashion in which virtually all of the arguments proceed. In orthodox science, one is expected at some point to go out and to do some research of one's own. Creation scientists rarely or never do this. Rather, they take the arguments, side comments and off-hand remarks of evolutionists, showing that in fact they support creationism rather than evolutionism. Everything is parasitic on the work of real scientists. Indeed, reading the major works of creation scientists (such as *Scientific Creationism* edited by Henry M. Morris, or, *Evolution: The Fossils Say No!* by Duane T. Gish) one would think that the main spokesmen for the fundamentalist cause are Ernst Mayr, G. G. Simpson, Theodosius Dobzhansky and Stephen Jay Gould! This, in itself, smacks of impoverished thought, verging on the dishonest. If Mayr and Gould are so reliable in their side thoughts, why are their main evolutionary conclusions to be ignored? Moreover, at times, creation scientists move right out into blatant fraud. For instance, one recent work, *Creation: The Facts of Life* by Gary E. Parker, refers constantly to "noted Harvard geneticist [Richard] Lewontin's" view that the hand and the eye are the best evidence of God's design.

Can this really be so? Has the distinguished author of *The*

Vintage Magazines

H. L. Mencken (left) made the citizens of Tennessee laughing stock with his reporting of the Scopes trial. Later, the trial was dramatised as the hugely successful Inherit The Wind. *The 1960 film starred Spencer Tracy (above left) as Darrow and Frederic March as Bryan*

Genetic Basis of Evolutionary Change really forsaken Darwin for Moses? Not quite. Looking at the actual context from which the quote was lifted (a brilliant article on adaptation in the September 1978 issue of *Scientific American*), we find that Lewontin says that *before Darwin,* people believed the hand and the eye to be the best evidence of God's design. *Now* we know them to be the products of evolution through natural selection. But you would never learn this from Parker's book, which makes Lewontin say the very opposite.

As one might expect, given their sources, most of the creation science arguments are less arguments for creationism and more arguments against evolutionism: the assumption is that if you are against the one, you are for the other. This, incidentally, was an assumption that got the Attorney General of Arkansas into hot water in the recent trial of the "balanced treatment" law's constitutionality. Astrophysicist Chandra Wickramasinghe is against orthodox evolutionism, and the Attorney General called him as a witness for the state. However, although he is certainly outside the mainstream of modern thought, Wickramasinghe is no Biblical creationist: he believes rather that Earth was seeded by life from outer space. On the witness stand, Wickramasinghe confessed that he could not see how anyone could hold to the views of the creationists, especially with regard to the supposedly short history of the Earth. The judge, like everyone else, was amazed at the damaging effect this had on the state's case.

Many of the creationists' anti-evolutionary arguments have a familiar ring. For instance, in virtually all of their works that old chestnut about natural selection being a tautology is dusted off and trotted out, with all sorts of citations to Sir Karl Popper and Sir Peter Medawar to back up its authority. This has been refuted many times, and Sir Karl Popper himself has even recanted in print. Then again, creationists constantly argue that a natural beginning to life is an impossibility, for such a process would violate the second law of thermodynamics. The coming of new life requires the natural production of order from disorder. The second law states that the process always goes the other way: you can scramble an egg but you cannot unscramble it. Therefore, life could never have appeared here on Earth, without the aid of the Great Designer in the sky.

No amount of pointing out the fallacious nature of this argument will budge the creationists. But, as any sixth-former ought to be able to tell you, the appeal to the second law in this context is quite irrelevant. The second law says nothing on the matter of new life, for the law applies only to closed systems. If one has the right kind of energy to do so, one can always make order out of disorder. The law states that in a closed (isolated) system, this usable energy will eventually be exhausted. However, this Earth of ours is an open system. Usable energy is coming in from the Sun all the time, and has been from the beginning of Earth's history. Hence, the possibility of order coming from non-order at some point here on Earth is in no way barred by thermodynamics.

The creationists spend a great deal of time on the fossil record. This in itself is a misleading tactic, for the fossil record has always been but one part of the evolutionary spectrum. Picking out the record for such extended discussion gives a false picture of evolutionary studies. In fact, evolutionists turn for insight and support to all areas of biology: biogeography, morphology, embryology, systematics and so forth. And yet, creationists hurry us through these fields, barely mentioning them, if at all. Consider, for instance, the finches of the Galapagos, which from Darwin on have always been a major brick in the evolutionary edifice. Creationists usually ignore them entirely—undoubtedly a deliberate omission, given the prominence of Darwin's finches in evolutionary writings. When they are mentioned, it is claimed that they are all the same species. A flat lie.

Most of the arguments based on the fossil record were devised even before *The Origin* was published, and not surprisingly, are somewhat dated. For instance, much is made of the supposed absence of Precambrian organisms. Why do trilobites appear in the fossil record full-blown, with no predecessors? Of course, the answer is that they do not. There is now massive Precambrian evidence, taking life back at least three-quarters of Earth's 4.6 thousand million year history. Creationists gloss over this.

Similarly, the creationist claim that gaps in the fossil record prove the impossibility of evolution no longer has any scientific validity. Bridging fossils have been found for all of the famous gaps. Take, for instance, the hiatus that existed between reptiles and birds. Even before the *Origin* appeared, a fossil feather was discovered that was shortly to be identified as belonging to *Archaeopteryx,* the organism that sits right on the fence between reptiles and birds. (Full specimens of *Archaeopteryx* were found early in the 1860s.) One could not hope for a more perfect example of a "missing link".

So, what do creationists do in the face of such evidence? They argue that *Archaeopteryx* had feathers (true), that evolutionists classify it as a bird (true), and that hence it is no link (false). Obviously, this line of argument is just a verbal sleight of hand. *Archaeopteryx* has feathers and a fused wishbone, like a bird. Like a reptile, it has a tail, teeth, small brain, separate digits, but lacks a developed sternum. It is right on the border, and has frequently been classified as a reptile by those ignorant of its feathers.

Perhaps the most incredible creationist argument concerns the roughly "progressive" nature of the fossil record. That one finds such a progression in the record, from relatively simple to relatively complex, was first noted at the beginning of the 19th century, by the French father of comparative anatomy, Baron Georges Cuvier. In *The Origin,* naturally, Darwin explained progression in terms of evolution up from humble forms, although by then all scientists recognised that the record is more than a simple progression: there is much branching through the course of ages.

Creationists cannot deny the record. They explain it all away as an artefact of Noah's Flood. Trilobites, being slow and sluggish, got trapped at the bottom by the rising waters. Dinosaurs managed to scramble half-way up, before the waters overtook and drowned them. Humans, most agile of all, got to the tops of hills before they met their watery fate (excluding any lucky enough to float away). Thus, we get the illusion of progression. But, say the creationists, it is no more than that: an illusion.

Comment is hardly needed. The creationists' discussions read like extracts from *Alice in Wonderland.* All we need to find are fossil remains of the Mock Turtle. (We already have the Dodo.) Just think! No trilobite was light enough, quick enough or lucky enough to swim to the top. No human was heavy enough, slow enough or sufficiently unlucky as to get trapped at the bottom. How improbable. Apart from anything else, knowing the average Englishman's fondness for his pets, do you really think that none would have taken his tame trilobites to the top? Or, that there would have been none so stubborn and bloodyminded as to stay in his cottage at the bottom of the valley, come hell or high water (or, as the creationists no doubt presume, come both)?

Even the creationists recognise the implausibility of their position, for they add that some human marks are found down the fossil record, along with the dinosaurs. But, no evolutionist need fear. The supposed "human" traces turn out to have a very fishy (or dinosaurish) look about them. Apart from anything else, the traces generally consist of "footprints" much larger than any human could make today. The lame excuse offered (as in *Genesis Flood* by John Whitcomb Jr and Henry Morris) is that in the olden days there were giants. But what authority is there for this? None other than our old friend, *Genesis* (6:4). The non-scientific circular notion of creation science is there for all to see.

Moving on to their final arguments, Creationists deal quickly with the age of the Earth. It is boiled down to around a millionth of what we learn from various radiometric dating techniques. (Think of it. The age of the Earth is a *million* times shorter than most believe.) How can this reduction be achieved? Quite simply. If one wishes to deny conclusions, then either one denies premises or one denies inferences. Creationists do both. For instance, they argue that one cannot assume that rates of radioactive decay are constant. The reasons why tend not to be specified too clearly.

In parallel with negative arguments against conventional dating techniques, the creationists for once offer a positive argument of their own. They suggest that if we extrapolate back from current human population numbers and growth rates, we arrive at a sole Adam and Eve around the time of God's original creative act (about 6000 years ago). If we accept evolution, assuming that *Homo sapiens* arrived here on Earth about a million years ago, then given population growth we would today have more humans than there are electrons in the world. An obvious impossibility.

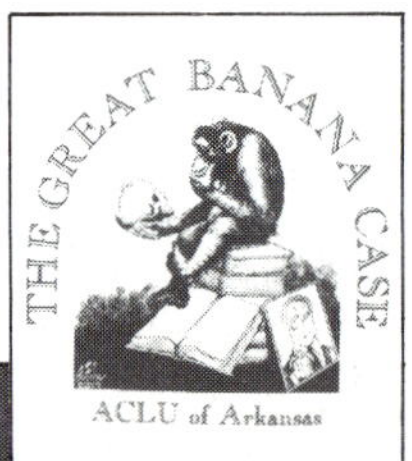

I hardly need say what a silly argument this is. The whole point of evolutionary theory is that population growth does not occur unchecked. Disease, famine, war, all lead to the Malthusian struggle for existence, which leads in turn to selection and evolution. Put matters this way. If there were no checks, we should be knee-deep in houseflies and herrings long before humans began to bother us. And in any case, the figures brandished by the creationists are phoney. They cite a human population growth rate of 2 per cent; but we have only started to edge up to this rate in recent years. Written records tell us that, in earlier times, growth rates were far lower than they are today.

Finally, there is the question that hovers over all discussions about origins. Not without reason was Darwin's theory known as the "monkey theory". What about us? What about *Homo sapiens?* Creationists flatly deny that the increasingly well-known hominid record has any relevance at all. There are no proto-human intermediaries. What about the much-feted *Australopithecus?* With respect, they were no human forefathers, for they had the brain of an ape and walked on their knuckles like an ape. Even *Homo erectus* is really no more than a rather run-down specimen of *Homo sapiens.*

If you believe this, you will believe anything. Suffice it to say that the recent discoveries of *Australopithecus afarensis* in Ethiopia put the total lie to the creationists. This small creature has an ape-size brain, and yet it walked around on its two hind legs, just like you and me. Even if one sat down and deliberately drew up a hypothetical ideal human ancestor, one could not do a better job than nature herself. Only by ignoring a massive amount of empirical evidence can one deny our simian ancestry (see chapter three).

Creation science is not science. It is crude dogmatic religion. For this reason, it is as offensive to the true believer as it is to the scientist. It can be maintained only through a perversion of our God-given reason. Interestingly, in their more candid moments, even the creationists admit the non-scientific nature of their case.

For instance, in *The Genesis Flood,* John C. Whitcomb Jr and Henry Morris write:

AP

Camera Press

Judge Overton (above) ruled in 1982 that Arkansas's creationism law was unconstitutional. But President Reagan endorses equal time for evolution and creation science in biology classes

> "But during the period of Creation, God was introducing order and organisation and energisation into the universe in a very high degree, even to life itself! *It is thus quite plain that the processes used by God in creation were utterly different from the processes which now operate in the universe!* The Creation was a unique period, entirely incommensurate with this present world. This is plainly emphasised and re-emphasised in the divine revelation which God has given us concerning Creation, which concludes with these words:
>
> > And the heavens and the earth were *finished*, and *all* the host of them. And on the seventh day God *finished* His work which He had made, and He *rested* on the seventh day from *all His work* which He had made. And God blessed the seventh day, and hallowed it; because that in it He *rested* from *all* his work which God had created and made.
>
> In view of these strong and repeated assertions, is it not the height or presumption for man to attempt to study Creation in

terms of present processes?" [pp 223-24. Their italics.]

I could not put the non-scientific nature of creation science better myself.

Incidentally, I am often asked why it was that the Attorney General of Arkansas never used the bigger guns in the creationist arsenal, such as Morris and Gish. It is suggested that he threw the case by not putting them on the stand. My hunch is that he realised he simply could not use them, for they have left such a trail of incriminating statements in their writings over the years.

Can you imagine a bright young lawyer cross-examining Morris on the above paragraphs? The Attorney General was forced to use people who had written virtually nothing. The one exception was Wickramasinghe, and look what a mess he made of the case for the defence.

One final point. The creation scientists like to portray themselves as martyrs. They claim that they do not want to exclude evolution. All they want is the right to have their own ideas brought before the world—or, more accurately, brought into the classroom. Hence, creationists argue that it is not they who are the bigots. It is the evolutionists. It is the established scientists who exclude ideas, insisting that no views other than their own may be taught in the classroom. But, conclude the creationists, this dogmatic approach obviously is contrary to the principles of sound education. Children should be exposed to all ideas, and then allowed to choose for themselves the one they prefer.

Will Hill

Nevertheless, smooth as it looks, this argument is perniciously fallacious. Sound education has never implied an indifferent purveying of wares before children, letting them pick and choose for themselves. The good teacher has always given to his or her children the truest and the best, as he or she knows it. A good physics teacher teaches astronomy not astrology, even though far more people look to the stars for their destiny than for confirmation of the Copernican revolution. Similarly, the good biology teacher teaches biology, not dogmatic religion. This is not dogmatic bigotry but professionalism.

Sound education is one thing. Teaching lunatic ideas to children is quite another. Creation science is plainly and unmistakably a fraud, and should not be taught alongside evolution.

Chapter Three

The evolution of man

Despite the lack of direct evidence, Darwin managed to produce ideas about the course of human evolution, some of which in retrospect look like inspired guesses. In other respects, however, he seems to have been wide of the mark

CHRIS STRINGER

CHARLES DARWIN was extremely evasive about discussing human evolution when he wrote *On the Origin of Species* . . . (1859) and he provided only one vague direct comment about the subject in the whole book. He discussed an important reason behind his reticence in the introduction to the book in which he did finally tackle the subject in detail, *The Descent of Man* (1871):

"During many years I collected notes on the origin or descent of man, without any intention of publishing on the subject, but rather with the determination not to publish, as I thought that I should thus only add to the prejudices against my views. It seemed to me sufficient to indicate, in the first edition of my *Origin of Species*, that by this work 'light would be thrown on the origin of man and his history'."

In 1871, only a few fossil apes and hominids (I use the latter term here in its narrow sense, to mean forms related to modern humans which date after the last ancestor shared with any of the modern apes) had been discovered, and only some of these are even mentioned by Darwin. For example, the Neanderthal discovery was still controversial in 1871, but Darwin devotes only 18 words to it, noting its large brain. He explained the situation thus:

VF

"With respect to the absence of fossil remains, serving to connect man with his ape-like progenitors, no one will lay much stress on this fact who reads Sir C. Lyell's discussion, where he shows that in all the vertebrate classes the discovery of fossil remains has been a very slow and fortuitous process. Nor should it be forgotten that those regions which are the most likely to afford remains connecting man with some extinct ape-like creature, have not as yet been searched by geologists."

So Darwin's work is based primarily on his observations of living humans and other animals, on their behaviour, comparative anatomy and development (embryology). From this study, he was able to make certain deductions about the likely course of human evolution and the factors which determined it. Some of these have been supported by subsequent research on living forms, and from fossil discoveries. But in other respects Darwin's guesses appear wide of the mark, judged from our present state of knowledge.

On our relationship to the apes, Darwin reported that Thomas Huxley had "conclusively shown that in every visible character man differs less from the higher apes, than these do from the lower . . . Primates". As to where the hominids arose, Darwin stated:

"In each great region of the world the living mammals are closely related to the extinct species of the same region. It is therefore probable that Africa was formerly inhabited by extinct apes closely allied to the gorilla and chimpanzee; and as these two species are now man's nearest allies, it is somewhat more probable that our early progenitors lived on the African continent than elsewhere."

The accumulating evidence of molecular and chromosomal studies amply confirms Darwin's own views on the very close affinity between human, gorilla and chimpanzee. Based on these approaches, the three forms appear to be about equally related to each other, while the orang-utan and gibbons fall at an increasing distance from them. If Huxley were alive today, surely even he would be astonished by data which suggest that we and the African apes differ in only about 1 or 2 per cent of our genetic material. We are more similar to each other than are species within many mammalian genera (such as dogs, cats and horses), yet man and the African apes are not usually classified within the same primate family, let alone genus.

Summing up the molecular and chromosomal data, there seems little doubt that the split between the apes (gibbons, orang-utan, chimpanzee and gorilla) and the human lineage was not a single event. Instead there was a series of splits, as first the gibbons, then the orang-utan, and finally the gorilla and chimpanzee lineages (we cannot resolve the details of this last event at present) diverged from the line leading to modern humans (see chapter eleven).

So while there was no single "missing link" between ape and human, there must nevertheless have been a common ancestor for the African apes and humans. Darwin made educated guesses about the form of the ancient hominid which must have evolved from that last common ancestor. In some respects these guesses have stood the test of time—and fossil discoveries—remarkably well. For example, Darwin was sure that man's ancestors walked upright before their brains began to develop. He said:

"Man alone has become a biped; and we can, I think, partly see how he has come to assume his correct attitude, which forms one of his most conspicuous characters. Man could not have attained his present dominant position in the world without the use of his hands, which are so admirably adapted to act in obedience to his will . . . But the hands and arms could hardly have become perfect enough to have manufactured weapons, or to have hurled stones and spears with a true aim, as long as they were habitually used for locomotion and for supporting the whole weight of the body, or, as before remarked, so long as they were especially fitted for climbing trees . . . From these causes alone it would have been an advantage to man to become a biped, but for many actions it is indispensable that the arms and whole upper part of the body should be free; and he must for this end stand firmly on his feet."

He had already argued that the development of the human brain should not assume exaggerated significance:

"We can further see why a great amount of modification in some one character ought not to lead us to separate widely any two organisms . . . Therefore in determining the position of man in the natural or genealogical system, the extreme development of his brain ought not to outweigh a multitude of resemblances in other less important or quite unimportant points."

The fossil evidence recovered from Africa during this century, and especially in the past 25 years, has established unequivocally that Darwin was right. In the evolution of the basic human specialisations, bipedalism preceded any substantial growth of the brain by at least two million years.

At Laetoli in Tanzania, remarkable evidence of the most direct kind about the existence of hominid bipedalism in the Pliocene has been recovered. By great good fortune, a series of ancient land surfaces made up of volcanic ash-falls has been preserved, then re-exposed. These floors record, through tracks and trails, the presence and movements of many species of vertebrates, including mammals and birds, and even invertebrates such as beetles. Also present are lots of tracks which can have been produced only by creatures who walked essentially as we do today.

Other deposits at Laetoli, of approximately the same or greater age, contain fossil vertebrates that can be matched with the preserved tracks. There are also fossil hominid remains, which Don Johanson and Tim White placed in the species *Australopithecus afarensis.* (In fact the Laetoli fossils include the type specimen of this newly-created and controversial species—see below.) The bipedal footprint trails must surely have been made by individuals of this species about 3.6 million years ago. *A. afarensis* clearly was capable of habitual bipedalism, which the skeletal material recovered from the other *afarensis* site at Hadar in Ethiopia confirms. This material includes the famous partial skeleton of a young female, nicknamed Lucy after the Beatles's song *Lucy In The Sky With Diamonds*. Some features of the Hadar fossils recall those of the apes, but Lucy's pelvis, and the various leg bones, confirm that bipedalism was well developed. It was not in some sort of semi-human state of evolution, it was clearly established, over three million years ago.

John Reader

Footprints at Laetoli in Tanzania. Animals walked across a fresh ash fall from a nearby volcano just before a rainfall which set the ash 3¾ million years ago. The animals included some fully upright bipeds

Why hominids should have evolved bipedalism at such an early stage in their evolution is still not clear. Darwin's emphasis on the advantages it gave for defence, attack and procuring food has been echoed by many anthropologists since. But clear records of tool-making and hunting behaviour appear in the archaeological record only at a later date. Owen Lovejoy, of Kent State University in Ohio, recently suggested that early hominids gained a reproductive advantage through the evolution of a monogamous mating system. Females remained at a home base and concentrated on rearing offspring, while their pair-bonded mates foraged for food. Those that walked upright liberated their arms for collecting food and carrying it back to the nuclear family at the home base. Various difficulties for this scheme have already been recognised, not the least of which is the fact that if the *A. afarensis* material actually represents a single species (and not all anthropologists accept that it does), its level of sexual dimorphism is higher than that found in any recent monogamous primate species. The males are much bigger than the females, which implies that the mating system of *A. afarensis* was polygynous rather than monogamous. However, the advantages of having free hands for collecting food must surely have played *some* part in establishing bipedal behaviour,

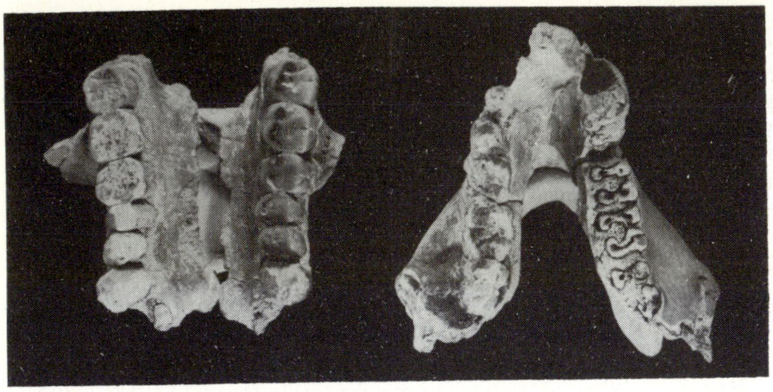

David Pilbeam

The jaws of Sivapithecus *(upper jaw at left). These almost certainly belong to a male who lived about nine million years ago. But the similarities to hominid teeth are probably misleading*

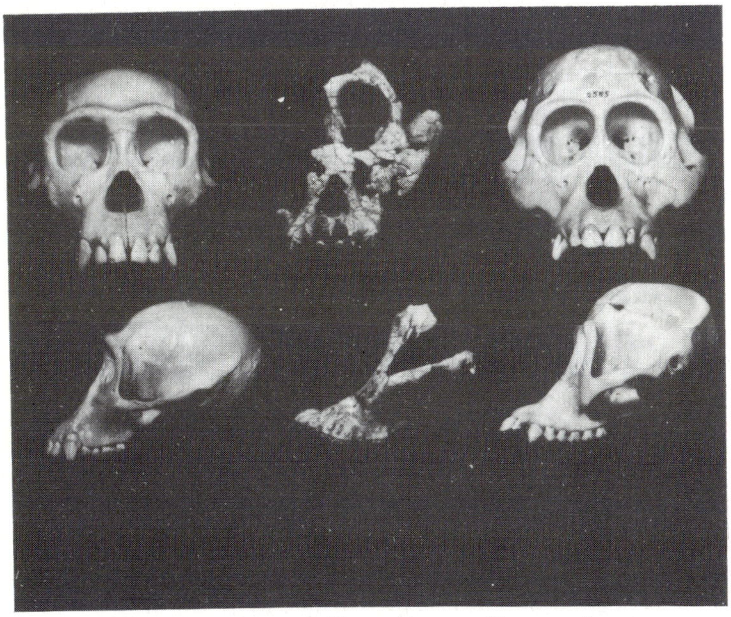

David Pilbeam

The new Sivapithecus *(centre) is much more like a modern orang-utan (right) than like a modern chimpanzee (left). It is not a hominid*

even if cultural activity became important only later.

On the evolution of the skull and jaws, Darwin wrote:

> "As the progenitors of man became more and more erect, with their hands and arms more and more modified for prehension and other purposes, with their feet and legs at the same time transformed for firm support and progression, endless other changes of structure would have become necessary. The pelvis would have to be broadened, the spine peculiarly curved, and the head fixed in an altered position, all which changes have been attained by man . . . As [our progenitors] gradually acquired the habit of using stones, clubs, or other weapons, for fighting with their enemies or rivals, they would use their jaws and teeth less and less. In this case, the jaws, together with the teeth, would become reduced in size, as we may feel almost sure from innumerable analogous cases [in other animals] . . . The adult skull would have come to resemble more and more that of existing man . . . As the various mental faculties gradually developed themselves, the brain would almost certainly become larger."

While it seems that Darwin was right about the precedence of bipedalism in hominid evolution, in the past few years there has been increasing doubt about the reasons he offered. This is especially true of the role of tool-use in establishing upright walking, and in reducing the size of the teeth, especially the canines. Only in the past 20 years or so have anthropologists managed to break out of the straitjacket of Darwin's reasoning (followed by Friedrich Engels and many others) that tool-use was a fundamental factor in the origin of hominid characteristics. First, we have learned that modern apes naturally can and do use and make tools in certain circumstances; yet this seems to have had little impact on their pattern of locomotion or on the shape of their teeth. Perhaps this is because they use tools so sporadically. But more significant than these observations on modern apes was the recognition that predominantly dietary factors might account for the dental characteristics of early hominids. There is no need to invoke the use of tools, or resort to the circularity of reasoning implicit in the feedback model that links bipedalism, hunting behaviour and tool making, canine reduction, and the evolution of the brain.

Bob Campbell

The famous 1470, the controversial skull that many people think is a very early Homo habilis

One pioneering model of this sort was developed in 1970 by Clifford Jolly, who argued that differences between typical savannah baboons and geladas paralleled in several respects those between apes and hominids. The differences between the two types of monkey could, according to Jolly, be attributed largely to dietary differences; geladas eat far more seeds in their diet. Jolly suggested that this distinction had led to the evolution in geladas of such characteristics as more dextrous fingers and unusual dental proportions. The front teeth, including the canines, are relatively small and the back teeth are large, with resistant surfaces to grind seeds effectively.

Jolly's "seed-eating hypothesis" has its imperfections, but it has led the way to a new model for the origin of hominid characters, based not just on a diet of seeds, but on one that required powerful grinding of abrasive foods such as roots, tubers and nuts as well. Such a diet may not seem appropriate for the teeth and jaws of modern humans, although we do have relatively smaller front teeth than the African apes, with larger back teeth and thicker enamel. But the explanatory power of the dietary model becomes especially relevant where Darwin's model fails.

When we turn to the fossil record we see—in the apes of the Middle Miocene, in *Gigantopithecus* of the Miocene-Pleistocene, and in the robust australopithecines (variously named *Australopithecus robustus, A. boisei,* or a separate genus, *Paranthropus*)—thick, robust jaws with relatively small front teeth and large molars with thick, protective enamel. Despite this thick enamel, the molars of these different fossils often show very heavy wear, which suggests a very abrasive diet. It seems that some of our dental characters may be due to an ancestral pattern established at least 10 million years ago in the Miocene, rather than to more recent tool use and meat eating.

This can help to explain a paradox that has puzzled anthropologists for many years. "Hominid" dental characters, like bipedalism, appear in the fossil record before good evidence of tool making; if the teeth were shaped by tool use, where is the evidence of the tools? The time gap (between "hominid" teeth and tools) is perhaps not as large as it seemed when some anthropologists, such as Louis Leakey and David Pilbeam, were induced about 15 years ago to propose that "hominid" dental characters present in the Miocene fossil ape *Ramapithecus* indicated incipient tool-making, meat-eating and bipedalism at that early date. But the gap is still a matter of more than a million years. (The interpretation of the dental characters in question has changed, and the hominid status of *Ramapithecus* is increasingly in doubt. Its Asian representatives, at least, seem to be closely related to, or even members of, the same genus as, the fossil ape *Sivapithecus.* And *Sivapithecus* now appears to lie on or near the ancestry of the orang-utan rather than that of the African apes and humans.)

So it is clear that African primates with hominid teeth were walking upright at least a million years before the first certain evidence of tool-making. One could of course argue that *A. afarensis* indulged in cultural activities that left no recognisable mark in the fossil record, and that these "invisible" activities shaped their teeth and legs. This may indeed be so, but the great dental resemblances between *A. afarensis* and the thick-enamelled apes of the later Miocene suggest that these similarities are the result of a common dietary adaptation to life in conditions more open than those in which the early Miocene apes lived. This move into more open country during the Miocene, which perhaps fortuitously produced some of the fundamental hominid adaptations, may have been due as much to ecological competition from the increasingly successful and diversifying forest monkeys as to any intrinsic "pioneering spirit" in our ancestors that made them exploit new resources.

Darwin said nothing concrete about possible rates of change in hominid evolution, but elsewhere he favoured a model of gradual change between species based on the accumulation of small modifications. In this respect he differed from T. H. Huxley, who was prepared to accept that evolution could operate through "jumps". In the past 10 years the model of punctuated equilibria has strongly challenged the Darwinian model of gradualism (see chapters seven and eight). The new model suggests that new species arise through a process completely different from the small changes that occur within a species. The transition between the new and the ancestral species would probably not be discovered from the known fossil record because speciation is so restricted in time and space.

Impressive support for the model of punctuated equilibria has recently come from Peter Williamson's study of molluscs preserved in the Plio-Pleistocene deposits of Koobi Fora in northern Kenya, where Richard Leakey has his major site. Species continue unchanged for hundreds of thousands of years and then there is a period of confusion, after which a new species emerges. But the relatively brief (in the relevant geological time scale) periods of developmental instability in these molluscs, as one species evolved from another, could be translated into perhaps 20 000 molluscan generations. In hominid terms this might be the equivalent of 400 000 years, but this spans the entire generally recognised period of the existence of *Homo habilis* or of our own species, *H. sapiens*. So in hominid evolution we are dealing with a really short time scale in geological terms, and gradualism or punctuation may well be in the eye of the beholder.

Last year I wrote a paper with my colleagues, suggesting that genetic data and the hominid fossil record did not support the punctuated equilibria model, despite claims to the contrary by such people as Stephen Jay Gould, Niles Eldredge and Steven Stanley. The examples that Gould and Eldredge chose from the fossil hominid record to support their case were certainly poor ones. This is not to say, however, that one cannot make a better case for punctuation and stasis in hominid evolution from other evidence. We attempted to demonstrate that members of *Homo habilis* were good intermediates between *Australopithecus africanus* and *Homo erectus*.

Despite this, some of the main characteristics of *Homo erectus*, which remain established in the fossil record for the next million years or so, seem to have appeared very rapidly. Similarly, Neanderthal characteristics, which had become established in Europe over a period of more than 100 000 years, vanished in less than 5000 years. We explained this by the view that the Neanderthals themselves were a relatively peripheral group of early *Homo sapiens*; gradualism elsewhere could have slowly established the "modern" traits, which then appeared rapidly in Europe through migration or gene flow between populations.

On present dating evidence, some populations of *H. erectus* (such as those in Indonesia) hardly changed at all, at least in the parts preserved, over a period of perhaps a million years. But other populations were not so static, and there are fossils from China, Europe and Africa which seem to represent mosaics between *H. erectus* and *H. sapiens* forms. Unless these are hybrids, or represent an unrecognised third Middle Pleistocene lineage of mosaics, they would appear to be good evidence of a gradual evolutionary change from *erectus*-like forms to *sapiens*-like forms over a major part of the Middle Pleistocene habitat range of humans.

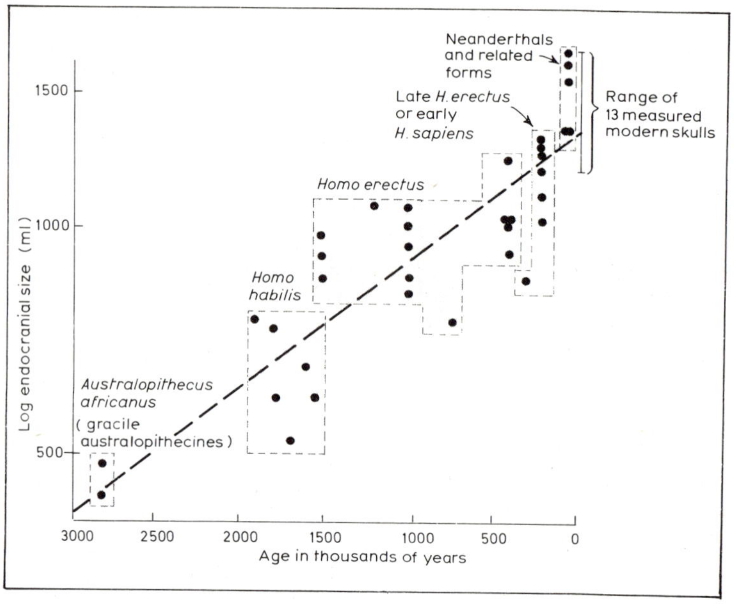

Measuring the rate of change in brain size during the evolution of man is no easy matter. In this graph I have relied on Ralph Holloway's data from casts of the inside of the skull to measure brain size, and where reliable dates are lacking I have selected dates on the basis of other fossils in the same beds. The result is a strong correlation (r = ·9) between brain volume and geological age. This corresponds to fairly consistent rate of increase of about 50 per cent per million years from the Pliocene 5 million years ago to the present day

Changes in brain size too, despite claims to the contrary, are explained better by the gradualistic model. Gradual increase in brain size during the Pleistocene still seems to hold good using the best available data and my own interpretation of dates for the relevant fossil material. This gradual trend could still be compromised, by more accurate dates for fossils or by other data concerning correlated changes in body size. An increase in body size over the same period could account for much of the increase in brain size, while if body size fluctuated in some unknown way a punctuation-stasis model could still fit the same data.

Given the limited and sometimes inaccurate data at his disposal, Darwin produced some inspired guesses about the course of human evolution. Although he seems to have been proved right about our very close relationship to the African apes, and the precedence of bipedalism over brain growth, he, along with many anthropologists, may have been wrong about the role of cultural behaviour during the early stages of hominid evolution. I have not had the space to discuss his views on the development of modern racial variation, but his emphasis on sexual selection rather than adaptation to climate as the mechanism seems also to have been wrong, although this is not to say that sexual selection has not played some part.

Finally, on the issues of rate and mode of hominid evolution in general, we must keep an open mind; I would not be too quick to write off completely Darwin's preferred model of gradual evolutionary change. Who knows, perhaps hominid evolution might turn out to be one of the exceptions that proves the rule!

PART II
IN THE BEGINNING

Natural selection applies not only to living things, but also to chemicals in the act of becoming living. More stable molecules, for example, and those that can use other molecules as a source of raw materials, will prosper. The transition from non-living chemicals to living organisms has been intensively studied, and answers that confirm Darwin's speculations are beginning to emerge

Chapter Four

Darwinism at the very beginning of life

Even before there was life there was natural selection. The biochemistry we see today is the outcome of those early struggles

LESLIE ORGEL

CHARLES DARWIN does not refer to the problem of the origins of life in the books he published during his lifetime, probably because he wanted to avoid unnecessary controversy. However, we know from one of his letters that he gave serious thought to the problem. The much quoted paragraph on the origins of life reads:

"It is often said that all conditions for the first production of a living organism are present, which would ever have been present. But if (and oh, what a big if) we could conceive in some warm little pond, with all sorts of ammonia and phosphoric salts, light, heat, electricity etc, present, that a protein compound was chemically formed ready to undergo still more complex changes, at the present day such matter would be instantly devoured or absorbed, which would not have been the case before living creatures were formed."

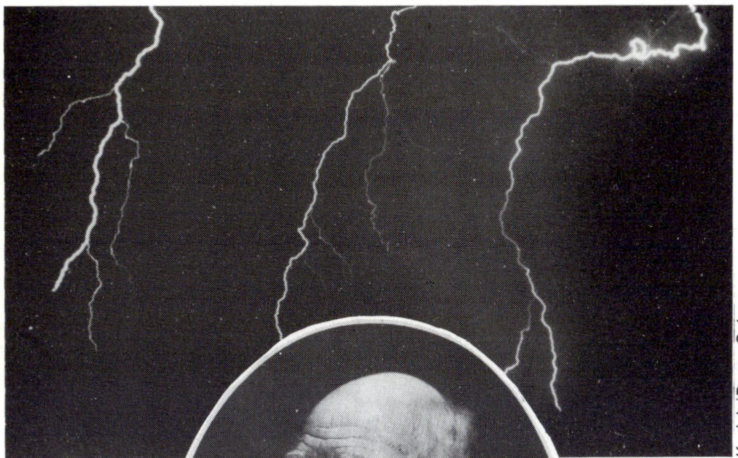

Kodak/Bruce Coleman

Darwin is arguing that we cannot expect to see life originating today because any suitable "prebiotic" medium would be eaten by living organisms, but he clearly believes that evolution of cells from inorganic constituents could have occurred in a pool on the primitive Earth. Modern views amplify but do not contradict Darwin's early speculations.

Recent studies of extremely old sedimentary rocks in Australia make it very probable that single-celled organisms, perhaps similar to blue-green algae, were already present on the Earth about 3.6 billion (thousand million) years ago. We know that the Earth accreted from lifeless dust and gases about 4.6 billion years ago. It seems, therefore, that life must have appeared on the Earth at some time during the first billion or so years of its history.

Theories that living organisms originated elsewhere in the Universe and were transported to the Earth in one way or another have been put forward from time to time. These theories cannot at present be tested, so they have had little influence on most thinking about the origins of life. In this chapter I will follow a conservative course and suppose that life evolved *de novo* on the primitive Earth.

It seems unlikely that we will find fossils or other geological relics of pre-life or the very earliest forms of life. Our ideas about the origins of life, therefore, are largely based on inference from our knowledge of contemporary biochemistry, on one hand, and laboratory reconstructions of the supposed chemistry of the primitive Earth, prebiotic chemistry, on the other. What do they tell us?

Recent work on the origins of life begins with the speculations of a Russian scientist, Aleksandr Oparin, who believed that the early atmosphere of the Earth, unlike its present atmosphere, was strongly reducing. He thought that the main components of the atmosphere were methane, ammonia and water, and that a pool of organic materials, a prebiotic soup, was formed from them by the action of ultraviolet light from the Sun, electrical storms, volcanoes, and so on. Oparin believed that life evolved in the prebiotic soup through a series of chemical reactions of increasing complexity. The demonstration that a prebiotic soup could have formed in much the way that Oparin suggested is the major achievement of origins of life studies. How life could have evolved from such a soup is the main question that remains to be answered.

Among the thousands of organic molecules that play some part in biochemistry, the 20 natural amino acids and the four or five nucleotides are particularly significant. The amino acids are the monomers from which the all-important proteins are constructed, while the nucleotides are the building blocks of the nucleic acids, which are the storehouses of genetic information.

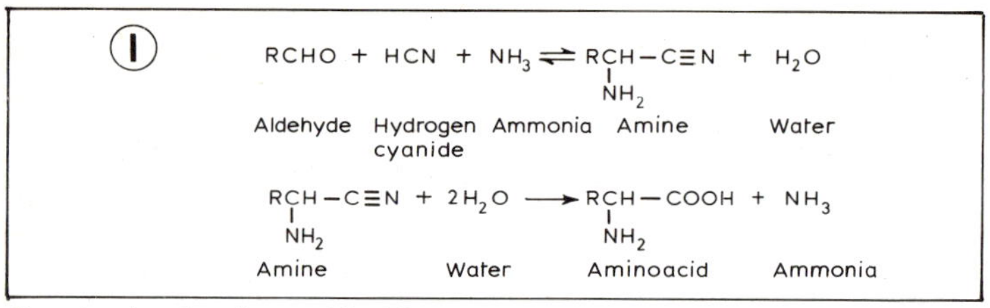

Synthesis of amino acids. The radical R *specifies the amino acid; hydrogen at that position makes glycine*

Stanley Miller and Harold Urey carried out the first modern experimental studies of the origins of life by demonstrating that amino acids could be synthesised by a non-living system. They subjected a mixture of methane, ammonia and water, Oparin's reducing atmosphere, to the action of an electrical discharge in a specially designed apparatus, which permitted the gases to circulate repeatedly through the discharge while non-volatile products were trapped. Miller and Urey claimed that the chemistry that occurred in their flask simulated the action of electrical storms on the atmosphere of the primitive Earth.

After the arc had been sparking for several days, Miller and Urey analysed the mixture of dissolved substances. About 20 per cent of the total carbon in the reaction mixture had been converted to a relatively small number of simple organic molecules, including substantial quantities of the simple amino acids glycine, alanine and aspartic acid. This result came as a surprise to organic chemists, because there are many thousands of organic molecules of comparable complexity and at the time there was no reason to expect anything other than an extremely complicated mixture of very many different compounds.

These successful experiments were followed by many other syntheses of important biochemicals from simple starting materials under prebiotic conditions, and also by studies designed to discover the mechanisms by which the biochemicals had formed. Miller showed that the glycine he obtained in his experiments was synthesised from formaldehyde and hydrogen cyanide that formed in the electric discharge. John Oro at Houston showed that hydrogen cyanide in the presence of ammonia would react to give adenine, one of the important components of nucleic acids. The remaining components of nucleic acids, the pyrimidines, were obtained in a similar manner from simple molecules like cyanoacetylene.

It soon became clear that the amino acids, sugars and nucleotide bases, the major components of the "genetic apparatus", can almost all be formed readily from a very small number of simple reactive gases that are formed whenever a reducing atmosphere is subjected to electric discharges, ultraviolet light or high temperatures. The most important of these reactive molecules are hydrogen cyanide, formaldehyde and cyanoacetylene which, together with ammonia and water, could be used to synthesise many of the most important components of proteins and nucleic acids.

The relevance of this work received strong support from an apparently unrelated area of science. Radioastronomers discovered that interstellar dust clouds, the regions where new stars are formed, contain very large amounts of organic material. Formaldehyde, hydrogen cyanide and cyanoacetylene are among the most abundant molecules. Furthermore, the chemical analysis of uncontaminated meteorites showed them to contain large quantities of the amino acids. All of this suggests that the synthesis of biochemicals suitable for the formation of a prebiotic soup must be occurring on a vast scale at many places in the Universe.

We do not yet understand in detail how the prebiotic soup on the primitive Earth was formed—there are many possibilities. Some compounds may have been present in the dust from which the Earth accreted, while others could have reached the Earth in meteorites and yet others formed in the Earth's primitive atmosphere by the action of lightning. While the details remain a matter of speculation, the general principle that important biochemicals are readily formed abiotically is well established. Amino acids and nucleotide bases are the major components of our genetic system in part *because* they were abundant in the prebiotic soup.

An everyday scene in a NASA laboratory (right) seeking to mimic the early Earth. The oldest known bacterium (below); it has already evolved a long way beyond the primeval soup

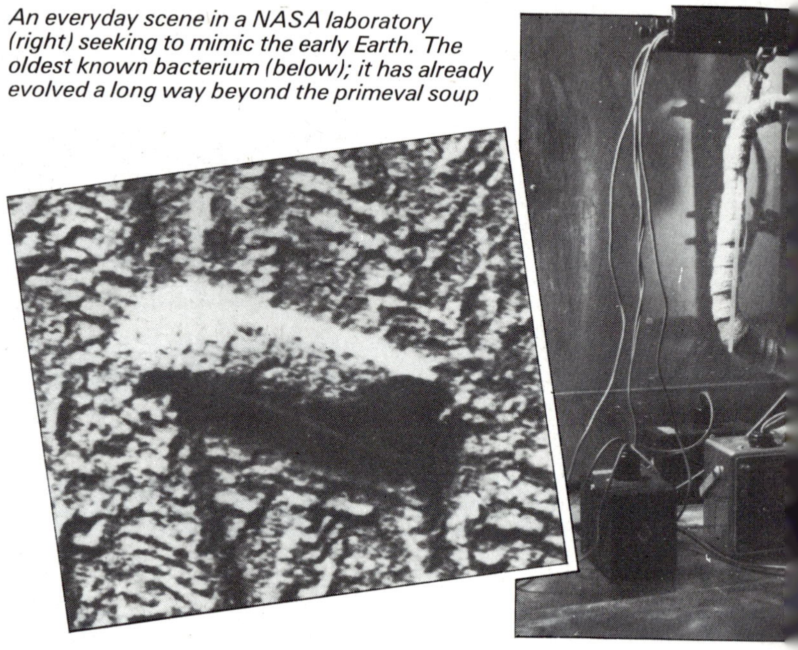

Prebiotic soup is easy to obtain. We must next explain how a prebiotic soup of organic molecules, including amino acids and the organic constituents of nucleotides evolved into a self-replicating organism. While some suggestive evidence has been obtained, I must admit that attempts to reconstruct this evolutionary process are extremely tentative.

Amino acids have been polymerised to give short, almost random, polypeptides in a number of ways. Sugars, nucleoside bases and phosphate can be combined to form the biologically important nucleotides under prebiotic conditions. In addition, random polymerisation of nucleotides to give short oligomers has been achieved experimentally. Thus relatively crude prebiotic syntheses of short random polymers resembling proteins and nucleic acids are already known. The major difficulties, intellectual and experimental, concern the biological organisation. How did molecular replication begin? How did the genetic code evolve? Were the first replicating systems surrounded by a membrane? If so, what was it made of?

Considerable progress has been made in the area of non-enzymatic replication. A preformed chain of uridylic acid residues, poly(U), will act as a template to enhance the formation of a chain of the complementary nucleotide adenylic acid (A). Similarly a preformed chain of cytidylic acid residues, poly(C), will facilitate the polymerisation of the complemen-

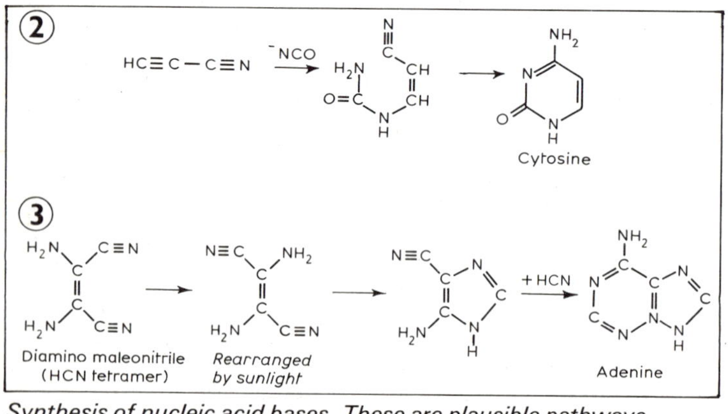

Synthesis of nucleic acid bases. These are plausible pathways for purines (2) and pyrimidines (3)

tary nucleotide guanylic acid (G). Such reactions always obey the Watson-Crick pairing rules; poly(U) has no effect on the polymerisation of G, and poly(C) has no influence on the polymerisation of A.

A few strikingly efficient and specific reactions have been described. The divalent zinc ion, Zn^{2+}, is a remarkably specific catalyst for the polymerisation of G on poly(C). Zinc promotes the synthesis of polymers of G up to 30 units long, while almost completely rejecting the other nucleotides. The divalent lead ion, Pb^{2+}, is even more efficient as a catalyst, but not nearly as specific—it produces a material that is very like a nucleic acid but is never found in nature. Very recently a template-directed system which faithfully and quite rapidly incorporates all four bases—without the help of enzymes—has been developed.

The ultimate objective of experiments of this kind is to develop a simple system, not dependent on biological enzymes, in which a polynucleotide of arbitrary sequence will facilitate the efficient synthesis of its complementary sequence. This would represent a major step towards understanding how nucleic-acid replication could have evolved, and

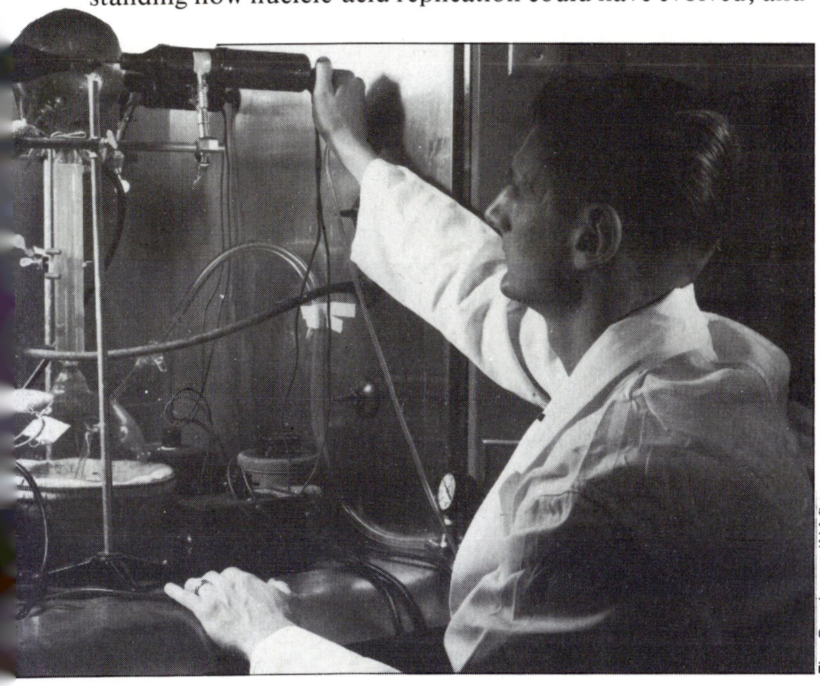

Else Barghoorn/NASA

we have made substantial progress in this direction.

Theoretical studies show that no great increase in biological complexity could occur until the nucleic-acid replication system became coupled to the synthesis of ordered polypeptides. The original coupling may, of course, have been far less precise than that which underlies the modern genetic code, but some primitive form of coding seems to be a necessity for the evolution of early life.

We do not yet understand even the general features of the origin of the genetic code. Theoretical hypotheses have been put forward which claim that the side-chains of the amino acids fit specifically into pockets in a polynucleotide structure formed in part by the codon, the anticodon, or some related sequence. None of these models is completely convincing, and none has received any experimental support. It is still unclear whether there was any direct stereochemical relation between amino acid side-chains and codon or anticodon structures. Francis Crick's "frozen accident" theory remains a viable alternative; the code we have today is probably the code that won an early Darwinian struggle. The origin of the genetic code is the most baffling aspect of the problem of the origins of life and a major conceptual or experimental breakthrough may be needed before we can make any substantial progress in understanding the code's beginnings.

Complex biological systems could not survive unless they had a method of holding together their constituent macromolecules. Nowadays, the cell membrane performs this function, so it is not surprising that a close analogue of the cell membrane has often been considered to be a necessary early feature in the development of cellular life. This view is not without its difficulties.

Modern cell membranes include channels and pumps which specifically control the influx and efflux of nutrients, waste products, metal ions and so on. These specialised channels involve highly specific proteins, molecules that could not have been present at the very beginning of the evolution of life. An impermeable membrane, without specific channels, would have been a disadvantage rather than an advantage early in the history of life, because it would have kept the useful components of the prebiotic medium outside and beyond the reach of the "cell's" machinery. It seems likely, therefore, that the macromolecular constituents of the earliest forms of biological organisation stayed together by some sort of self-aggregation, perhaps stuck to mineral surfaces, in a form that permitted ready access to nutrients in the "external environment". The development of a continuous membrane probably occurred relatively late, after complex metabolic pathways had evolved. Once a self-sufficient system of nucleic acids and proteins enclosed in an impermeable membrane had evolved we enter the realm of evolutionary biology and all that that entails.

Since the time of Louis Pasteur, the origin of optical activity in biological systems has attracted a great deal of attention. Two very different questions must be answered. First, why do all amino acids in proteins or all nucleotides in nucleic acids have the same handedness? Secondly, why are the amino acids all left-handed (L-) and the nucleotides all right-handed (D-)? We do not know the answer to either question, but we can make a number of plausible suggestions.

Many of the important structural elements that occur in proteins and nucleic acids cannot be built from a mixture of L- and D-isomers. A right-handed DNA structure can be built with D-nucleotides, and a mirror-image left-handed structure can be built with L-nucleotides. On the other hand, a mixture of D- and L-nucleotides would create a completely different structure or no regular structure at all. Thus if nucleic acid replication was the first evolutionary process, handedness may have been decided from the beginning. Alternative explanations of handedness depend on regularities in protein structure, the convenience of a unique handedness in simplifying metabolic conversions, and so on.

Theories of this kind do not explain why the system of D-nucleotides and L-amino acids was chosen rather than the mirror image system. Here there are two competing theories. According to the first, some asymmetry was present on the primitive Earth that strongly biased the choice. The ß-decay of radioactive elements such as potassium generates optically polarised electrons. Perhaps these destroyed the D-amino acids preferentially, thus favouring the L-amino acids. A number of other explanations of this general kind have been suggested. It is by no means clear, however, that any of the effects proposed is large enough to have biased the outcome significantly. The alternative view, and the one which I favour, is that the success of the D-nucleotide, L-amino acid system was an accident. The first great biological struggle set the D-polynucleotides in an equal contest against the L-polynucleotides, and the D-system won.

If we look back at the passage from Darwin's letter, we find that nothing that has been discovered since it was written contradicts his speculations. In fact, a major conceptual advance of the past several decades is the realisation that natural selection in the Darwinian sense must have begun at the period of molecular evolution, long before the appearance of the first "modern" organisms. Had Darwin had access to present-day knowledge of the chemical nature of the genetic system, I suspect he would have considered this conclusion obvious.

Chapter Five

Cradles for molecular evolution

The three-dimensional conformations of nucleic acid double helices may have played a key role in organising components from the 'prebiotic soup' into the ancestors of proteins and more advanced nucleic acids

CHARLES CARTER

WHAT structural features distinguished the first "biological" polymers from other products of random assembly processes? Why were they able to begin to evolve? Did they resemble in any way any of the components, such as ribosomes or polymerases, of present day molecular biology? I should like to describe a recent model-building experiment which suggests that the first self-reproducing system may have consisted of short polypeptides and polynucleotides which could form complementary secondary structures very much like those presently found as parts of protein and transfer RNA molecules.

I came to do this experiment with Professor Joseph Kraut at the University of California at San Diego by a somewhat indirect route. So I shall first describe some of the ideas and the clues which suggested to us that these rather formidable questions might be approached by model building.

The primitive Earth's chemistry is widely believed to have created a "prebiotic soup" containing the building blocks—amino acids, purine and pyrimidine bases, and sugars—for proteins and nucleic acids. Since these conditions no longer occur on Earth, they must be recreated in the laboratory. To study events which ensued in the prebiotic soup, and which eventually produced living cells, we therefore have to guess which materials should go in our test-tubes.

One can easily imagine how the building blocks might have been joined to form random polymers, but it is very much of a mystery how any of these products came to be selected for and so began to evolve. The problem is that evolution as we tend to think of it requires two very different properties: evolving molecules must function in some advantageous way, and the structure responsible for this function must be specified by some encoding process. The simultaneous requirement for function and coding suggests that even the earliest antecedents of present day proteins and nucleic acids were self-organising systems of molecules having complementary template surfaces. Such systems combine functional and encoding properties: the function of the template being to assemble its complement.

The structural requirements of such a system seem to pose substantial obstacles to primitive Earth chemistry. Present-day proteins can function primarily because they derive stability and specific binding pockets from the fact that their long chains can coil back upon themselves. Tertiary interactions of this sort seem too sophisticated to have been within the modest synthetic capability of the prebiotic soup.

One way to simplify this problem is to assume that primitive biopolymers could recognise one another by forming complementary secondary structures. The simplest secondary structure for a polymer is some form of helix, in which a constant relationship between a residue and its neighbours is maintained along the backbone. Helices may be constructed almost without regard to the sequences of their side chains, so that a very large number of different sequences might perform the same function. Furthermore, biological polymers, such as polypeptides and polynucleotides, can form helices which are stabilised by internal hydrogen bonds. This property is particularly advantageous because such helices can remain folded long enough to serve as template surfaces.

Joseph Kraut and I had been studying three-dimensional structures of various proteins with the idea that perhaps a

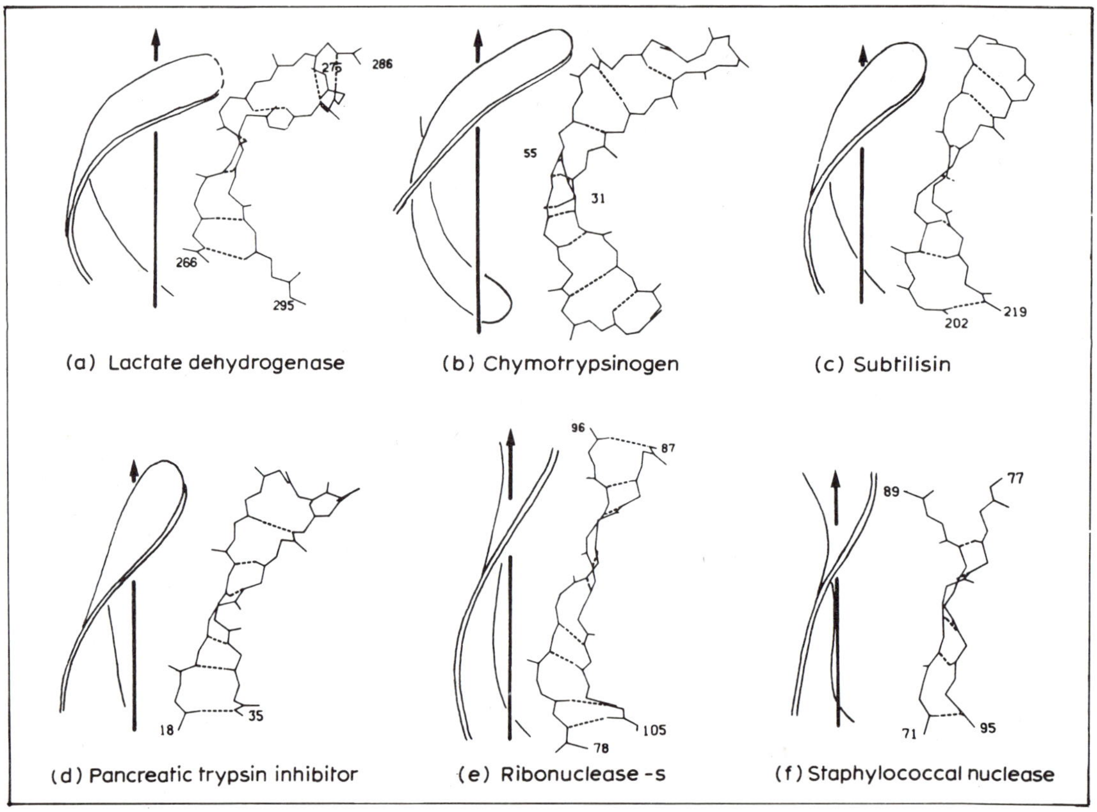

Figure 1 *Twisted ß sheets from six known protein structures. Each structure is formed by folding the polypeptide back on itself so that adjacent chains run antiparallel, that is in opposite directions. The ribbons drawn to the left of each structure are meant to suggest relative differences between the approximate pitch (the advance along the helix axis for a complete turn) and radius (how far away the popypeptide is from the vertical helix axis) for each structure*

polypeptide secondary structure could be complementary to a nucleic acid double helix, thereby eliminating the requirement for tertiary structures by providing a simpler stereochemical basis for precellular protein-to-nucleic acid binding.

An essential clue that such a polypeptide structure might exist emerged from an observation, made independently by Cyrus Chothia of the Molecular Biology Laboratory at Cambridge and myself, about "twisted" ß sheet structure in proteins. Individual polypeptide strands in a ß sheet always coil round one another to form a twist. In principle, there are two possible senses for such a twist. In proteins, however, it surprisingly turns out that both parallel and antiparallel structures always twist with the sense of a righthanded screw. This is the same sense as the twist in RNA and DNA double helices. Moreover, Chothia has shown that biological polypeptides, because they contain only L-amino acids, will actually *prefer* a righthanded twist when they form ß sheets. In contrast, a ß sheet made entirely from D-amino acids would be expected to twist in the opposite direction.

I compared sheet segments from nine of the known protein structures by making computer-generated drawings from atomic coordinates, many of which were supplied by other protein crystallographers. Six of these drawings are shown in Figure 1. Two aspects of these drawings caught my attention. First, each structure is actually an antiparallel double helix of polypeptide. They are therefore formally very similar to nucleic acid double helices, which are antiparallel pairs of polynucleotide chains. Secondly, and perhaps more remarkable, two of the six structures, **a** and **b**, have approximate pitches and radii very close to those found in double stranded DNA and RNA molecules. This highly suggestive discovery electrified us with the following thought: perhaps there is a fundamental structural complementarity between polypeptide and polynucleotide double helices.

To give this idea a preliminary test, we decided to find out just how precisely polypeptides and polynucleotides could be fitted together in this manner. So we constructed a skeletal model of a complex in which a polypeptide double helix ran along the groove of a nucleic acid double helix. We wished to construct and study as convincing a model as possible, and for this reason we decided to examine in detail only one of many possible structural combinations.

First, we fitted the polypeptide double helix into the minor groove of an RNA double helix. (The two strands of a double helix form two sorts of grooves (Figure 2), a wide major groove and a narrow minor groove.) We had compared the width and hydrogen bonding possibilities of the major and minor grooves of both RNA and DNA helices. The minor groove in RNA is wide enough to accommodate two antiparallel strands of polypeptide, and it is the only one of the four grooves offering the opportunity to donate a hydrogen bond to the exposed peptide carbonyl oxygen atoms. The ribose 2′ hydroxyl group, which we expected might participate in this hydrogen bond, is the one that is missing from the deoxyribose phosphate backbone of DNA molecules.

Secondly, we built the complex with adjacent polypeptide and polynucleotide chains running parallel, that is with peptide linkage in the N-C direction associated with nucleotide linkages in the 5′-3′ direction. This choice was dictated by the fact that when adjacent chains ran antiparallel the polypeptide was actually too wide to fit comfortably into the groove. When a double stranded ribbon of polypeptide is wrapped around a cylinder such as an RNA molecule, alternate pairs of ß carbon atoms point inward while the remaining ones point outward. One of these sets of ß carbon atoms is actually about 1 angstrom farther apart than the other, and we found that it was best to have this wider set pointing away from the RNA helix.

Considerations such as these specified enough features of the complex we wanted to construct to convince us that with small adjustments we could fit the polypeptide and polynucleotide double helices together with great precision. We made the necessary changes by a series of successive approximations. Then we used computerised techniques to check that in making these adjustments we had not unduly bent or stretched any of the parts. The resulting model is represented in Figure 3. It shows rather convincingly that a hand-in-glove fit can occur between polypeptides and polynucleotides.

I have already suggested that polypeptides and polynucleotides in solution will, for thermodynamic reasons, tend to assume conformations which favour the interaction shown in Figure 3. This conclusion rests both on theoretical considerations and on the fact that we can observe similar structures inside protein and transfer RNA molecules. It is thus one of the most important reasons why we believe that the model complex could indeed represent a real structure.

A closer inspection of Figure 3 reveals two striking coincidences which reinforced our conviction that the complementarity evident in the complex might have played an important role in precellular evolution.

First, the two double helices have exactly the same pitch and number of residues per turn. This means that they are synchronised in the sense that the same stereochemical relationships between them can be maintained throughout an arbitrarily long helix. Apparently, this synchrony can occur only when the two double helices are fitted snugly together. If they were closer together they would repel each other, and if they were farther apart unfavourable empty spaces would occur between them. In either case, one of the two helices would require fractionally more residues than the other, in order to complete a turn. Remarkably, RNA and polypeptide double helices are just the right distance apart if and only if they have the same number of residues per turn.

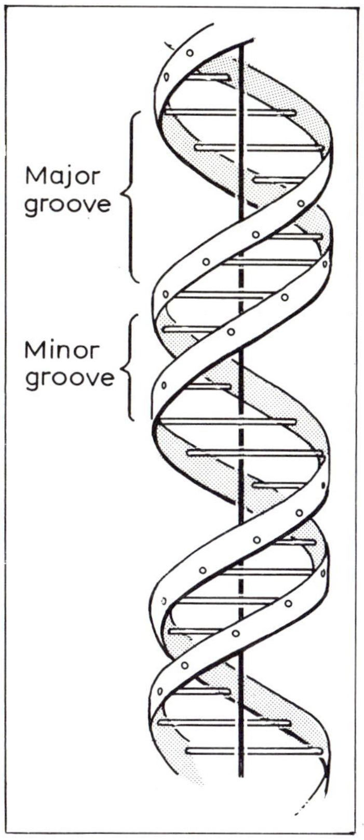

Figure 2 *DNA double helix showing major and minor grooves into which protein chains could be expected to fit. In fact, they fit neatly only in the minor groove*

Secondly, a set of repeating hydrogen bonds between the respective backbone chains prevents the ribbon of polypeptide from sliding along the groove in the RNA helix. The 2′ hydroxyl groups which project into the minor groove are positioned in such a way that they can donate their hydrogen atoms to free peptide carbonyl oxygen atoms forming hydrogen bonds in the manner shown by the heavy dashed lines in **a** of Figure 3. The relative position of adjacent RNA and polypeptide backbone chains, determined with considerable specificity by these hydrogen bonds, coincides with a particularly favourable juxtaposition of the RNA base pairs and the inward facing amino acid side chains, as is shown in **b** of Figure 3. Inward facing side chains occur between successive base pairs, so there is room for a ß carbon atom near three of the four bases.

Guanine, however, has an extra amino group which occupies the space normally taken by the ß carbon atom. In this one instance, glycine would be a favoured amino acid. Moreover, if the model is constructed with the appropriate inward facing side chains—alanine where a methyl group is

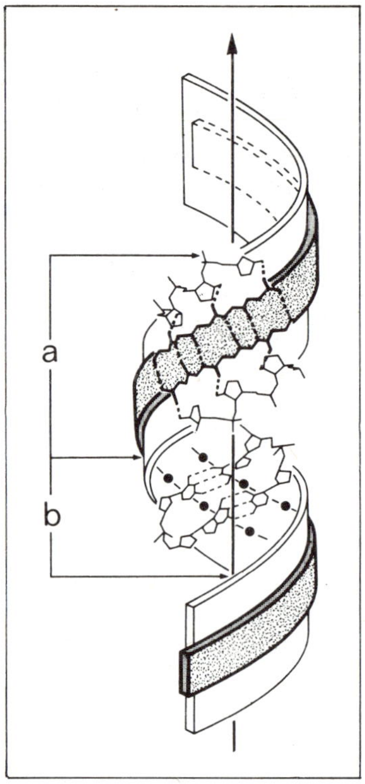

Figure 3 *The RNA-polypeptide complex. The polypeptide (shaded ribbon) wraps around in the minor groove of an RNA helix. The backbone chains are fixed relative to one another by specific hydrogen bonds shown as heavy dashed lines (a). Alternate amino acid side chains point inwards toward the RNA base pairs (b)*

allowed, and glycine where it is forbidden—there are still empty regions that have room for additional side chain carbon atoms. These voids would tend to be filled, in any real structure, by small amino acid side chains or by aromatic side chains capable of intercalating between successive base pairs. In this way we can envisage ways to achieve more detailed complementarity and perhaps even a primitive form of genetic coding between specific polypeptide and polynucleotide sequences.

It is hard to dismiss as coincidental the superb complementarity evident in our model complex. On the contrary, the interactions I have described provide so convincing a basis for matching polypeptides with polynucleotides that we have proposed that, in addition to being structurally complementary, each component separately can assemble copies of its complement from suitably activated precursors in the "prebiotic soup", thereby acting as primordial polymerases. Over a geological time span, even a very slow catalytic cycle of polymerisation, dissociation and subsequent polymerisation would cause the accumulation of different "ribonucleoprotein complexes" with varying degrees of stability and catalytic activities. Gradually, populations of the more stable and/or more prolific complexes would grow at the expense of those less well-endowed. In this way evolution of biopolymer sequences by a natural selection process could have begun in the prebiotic soup with a minimum of improbable preconditions.

I want now to think a bit more generally about the origin of template-directed biosynthesis. The most important element of our proposal is the fundamental geometrical complementarity between polypeptide and polynucleotide backbone chains. I have tried earlier to suggest that a very large number of different sequences would be compatible with the fit between main chains, and that this freedom from rigid sequence restrictions would have been advantageous to primitive self-replicating systems. This means that the earliest distinctively *biological* molecular properties would not have been specific sequences of amino acids or bases, but rather, the chemical linkages that promote backbone chain complementarity. Each successive linkage must, of course, be identical. Otherwise the chains cannot form regular helices. But there are many different ways to put amino acids or ribonucleotides together to make chains. In fact, when we try to link them together in a test-tube we get rather complicated mixtures of different linkages, and often the non-biological linkages predominate. Our proposal provides a plausible explanation for how the correct linkages came to be made. If the relationships shown in Figure 3 remained crucial throughout subsequent evolution, it also explains why only these linkages are found in proteins and nucleic acids today.

I have consistently assumed that the earliest biologically significant self-replicating system must have involved both RNA-like and protein-like components. I find much less probable the alternative assumption, that is that ancestors of the elaborate complexes of RNA and proteins present at virtually every stage of present-day molecular biology were created by a completely different molecular system, no trace of which remains today. The central problem of the origin of life seems to me to be how the flow of information from nucleic acid sequences into protein sequences developed. I doubt that any nucleic acid structure alone, even if it could replicate by virtue of complementary base pairing, could evolve the structures necessary to translate the triplet code into protein sequences. However, given assistance by polypeptides in the manner I describe, it is possible.

I think that it is quite hopeful that our model provides a way to make the correct chemical linkage in both polypeptide and polynucleotide chains. Surely, base triplet codons can be recognised as such only if they are properly arranged in space, and this arrangement depends upon having the correct linkages between them. However, this system has little else in common with the present genetic code, which uses three successive bases to determine one amino acid. Our model has two amino acids for each nucleotide base, so it can specify only half of the amino acids in a polypeptide, namely those which face the RNA base pairs. Nevertheless, we can glimpse in a general way how the accumulation of short ribonucleotide complexes such as that shown in Figure 3 might intensify this modest trickle of information and begin to exploit the enormous potential of folded, globular proteins. For example, we can imagine, in addition to the "polymerase" activities we have already proposed, other mechanisms which would add to or stitch together existing segments. These mechanisms might operate whenever dissociated components happened to rejoin out of phase, or somewhat unscrewed with respect to one another. If by this mechanism two identical segments became stitched together in an inverted configuration, the products would be double helices whose individual chains both had internally complementary sequences. These molecules could dissociate from their homologous partners and fold-up upon themselves forming "hairpin" loops. In this way the primordial population could grow in length and complexity, as well as in numbers, perhaps giving rise eventually to molecules with tertiary structure. Thus we can at least speculate on how subsequent evolution of our primitive system might generate molecules sophisticated enough to learn to read the triplet code.

Does the model complex shown in Figure 3 actually represent a real mode of interaction between polypeptides and RNA, or does it simply reflect a set of arcane coincidences? There are two different ways in which our proposals might be verified or rejected. First, we can hope to find relics of the complex incorporated into present-day proteins and nucleic acids. There are numerous occasions for such interactions in the ribosomes and polymerases involved in protein and nucleic acid biosynthesis. Many of these molecules are very complicated, but structures are already being worked out for transfer RNA molecules and for several of the enzymes which add amino acids onto transfer RNA. The second, and more convincing, proof of the validity of our proposals would of course be to demonstrate in a test-tube the binding and catalytic properties we have predicted.

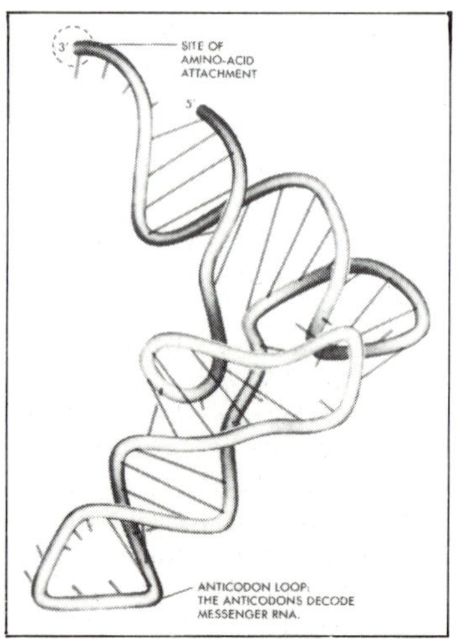

Three-dimensional structure of transfer RNA. This might have evolved in association with the proteins that it helps to build

PART III
PROCESS IN EVOLUTION

How do species evolve? H. B. D. Kettlewell provided beautiful evidence of adaptation, taken by Darwin and the neodarwinists to be the basis of species differences. Challenges to this view emphasise other processes that could mould the outcome of evolutionary change. The transition from reptiles to mammals is clear enough, but the underlying processes are still not certain

Chapter Six

Evolution and the environment

The last century saw an example of evolutionary change which is the most striking on scientific record. By the phenomenon called 'industrial melanism', whole populations of moths changed from white to black in as little as 50 years

H. B. D. KETTLEWELL

THE environment in the past has nearly always been governed by the physical conditions imposed by the weather, continental drifts, sun spots and other natural phenomena. These have, for the most part, brought about gradual changes over periods frequently measured in thousands of years. Animals and plants have therefore had time to become adapted to slowly changing circumstances. Modern concept believes that natural selection, working on the practically unlimited inheritable variables, has been responsible for living creatures adapting and surviving.

Ardea

The peppered moth, clear evidence of natural selection in action. As tree-trunks became blackened, the dark moths were selected

The advent of the Industrial Revolution brought about profound environmental changes which were man-made, and affected hundreds of square kilometres of the Earth's surface in a comparatively short period. It took place insidiously, and its implications and magnitude were entirely disregarded. The millions of tonnes of coal which each year go up the chimneys were indeed recognised as being the cause of the pollution that fell around the industrial cities of this country, north Europe, America and elsewhere, but it is not generally realised, even yet, that the amount is so great that it is measured in tonnes per square kilometre per month, and that 50 tonnes or more per month is sometimes recorded. Its cumulative effect as it is precipitated night and day, year after year, over vast areas is indeed profound.

These are the changes which ensue: leaves are contaminated and it therefore follows that the later in the summer we examine deciduous leaves, the dirtier they will be. It is not surprising then that evergreens are the dirtiest of any. Green fly secrete their sticky "honeydew" which serves to hold the pollution, apart from the copious number of sooty moulds which grow on it. Tree canopies are constantly being washed by rain, and all trees have well-marked "rain runs" commencing beneath the boughs and continuing down the trunks. There are considerable concentrations of pollution in these areas: the vegetative lichens are killed, the trunks and boughs become bare and ultimately black. In certain districts such as the industrial areas of Yorkshire the twigs, rocks and even the ground itself become black.

All these are the normal resting places of many species of night-flying moths who have to pass the day sitting exposed, but camouflaged by their "cryptic coloration", so that they are hidden from the predators which normally feed on them. Many of their original patterns were highly specialised for resting on lichens and were "near-perfect", as the photographs of *Biston betularia*, the peppered moth, show. The first records of the black form (*carbonaria*) of this species were made about the middle of the last century (1848) and were confined to Manchester. *Carbonaria* was found somewhat later in the industrial areas of Yorkshire and the adjoining counties. They were adjudged at that time as great rarities, and were bought by collectors.

Today, one hundred years later*, *carbonaria*, the black form, constitutes 90 per cent of the population in all industrial areas and, what is more interesting, far beyond. The frequency map, which we built up on more than 20 000 records from 170 observers, shows that a high proportion of *carbonaria* occurs throughout eastern England, and as far as the coast—which is a long way removed from the industrial centres but not from their effects. There is every reason to believe that this situation reflects the indirect effects of long-standing pollution, carried from urban areas by the prevailing westerly winds. Throughout the eastern counties, although lichens still remain on the trunks of such trees as apple and ash (probably in greatly reduced quantities), black "rain-runs" can nearly always be seen beneath the boughs, particularly of beech and oak, and these are the normal resting places of *B. betularia*.

*Dr H. B. D. Kettlewell was writing in 1958

The frequency map of the peppered moth in Great Britain, built up on more than 20 000 records from 170 observers, showing how a high proportion of the black form (carbonaria) *occurred in the 1950s throughout eastern England. In the 1840s it was confined to the Manchester area. The area within a circle represents, open, the proportion of the light form; solid, the proportion of the dark form. (The striped area is another dark form,* insularia*)*

Scientific American

Furthermore, no fewer than about 70 other species of moth are in the process of changing their populations from their old complicated light patterns to one of all-black coloration. This occurs only among those species which depend on camouflage, and have to sit exposed by day, but not among those species which have developed other forms of protection, such as warning coloration. The black forms are always inherited according to the laws of Mendel and are usually dominant. The change from white to black is a one-step difference and but rarely are there intermediates.

Now, recent work has been concentrated on ascertaining why and how this rapid spread has taken place. Many biologists found it difficult to accept that such profound changes could be brought about by natural selection in so short a time. In any case, we had absolutely no evidence of birds or other predators finding and eating cryptically hidden moths. New theories were evoked to account for the rapid spread of species changing from light coloration to black—the "industrial melanics"—and, but a few years ago, it was considered possible that it was the direct effect of some substance (a "melanogen") in air pollution on the moths themselves, and that this was inherited.

In view of the fact that there was no evidence that birds and other predators took and ate resting moths by day, and certainly none that they did this selectively, we decided to put it to the test by large-scale release experiments. Several thousand *betularia* of both forms were bred and, after being marked on the underside of a wing by a spot of quick-drying cellulose paint, whose colour was varied according to the day, they were released on to the trunks and boughs of trees, which is their normal resting place.

In the first instance, we did this in an unpolluted wood in Dorset where lichens grew thickly, and where the black form does not occur. This type of wood is probably similar to what the majority of woods would have looked like 200 years ago. In a release of 969 individuals, we recaptured twice as many *typica,* the light form, as *carbonaria.* We repeated this experiment in a heavily polluted oak wood near Birmingham, where *typica,* the light form, is rare. We freed 584 moths of both forms, but here we obtained a complete reversal and got back approximately twice as many *carbonaria* as *typica.* Furthermore, we were able by direct observation through fieldglasses to record what happened and the order in which the moths were eliminated. In the first wood, the nuthatch, spotted flycatcher, robin, yellow hammer and thrush were seen to take and eat the moths. In the second, we saw the robin, hedge sparrow and redstart.

These findings showed that the birds *took the moths selectively,* and provided evidence that the differences in cryptic coloration alone could be responsible for the rapid spread of industrial melanism. It appears that the black form has an approximate 30 per cent advantage over the light form in industrial areas. If we accept this figure, and also accept that in 1848, the date of its first capture in Manchester, not more than one per cent of the population there (let us say one million) was black, we are able to assess very roughly the following facts: that the original mutation, assuming that there was only one, which is most unlikely, would have taken place about 1810, approximately 50 years after the commencement of the Industrial Revolution; that by 1839 it would have reached the one in a thousand level and, as stated, by 1848 one per cent. Thereafter, the black mutant would spread at a very great speed until it formed 50 per cent of the population, and after this point at a slightly slower speed. In actual fact by 1900 *typica* was but rarely seen in Manchester. The same has happened throughout industrial Britain.

It is unlikely, however high the selective pressure, that the light form will entirely disappear because of the advantage of the heterozygote (a moth with one gene carrying black and one carrying white) over the homozygote black (with both genes black). Therefore, it is probable that in these areas the white form will continue to occur in the population at a frequency of 2 to 5 per cent, and that this will eventually become a balanced polymorphism.

Although *betularia* has been investigated more fully than any other species, there is ample evidence that about 70 other species are behaving in a similar way. It is probable, however, that some species mutate to their melanic forms more rarely than others; thus the dark form of *Ectropis consonaria,* a species which has been driven out of most urban areas in its normal light form, has in this country mutated successfully only on two occasions near Maidstone and near Stroud. In Germany, however, it has become an industrial melanic. In much the same manner the black form of *Colocasia coryli* has spread from what was probably a single centre in the Chilterns.

So far I have discussed colour differences only, black and white. But we have, apart from this, accumulated evidence that the same gene mutation which confers the changed line of pigment chemistry from white to black is also responsible, directly or indirectly, for other far-reaching effects in the same individual. Without doubt there are physiological and behaviour differences; the larvae of the black and the white form (of industrial origin) feed at different speeds; black and white virgin females appear to be attractive to typical males to a degree which varies with the temperature; the present-day black form *carbonaria* is hardier in the face of disadvantages than *typica*; there is also evidence that the two forms are capable of choosing backgrounds that conform with their own colours on which to pass the day.

What is more important is that the presence of the *carbonaria* gene at a high frequency in the population has been followed by adjustments throughout the whole of the rest of the creatures' gene-complex. Modifying genes, which previously have conferred a disadvantage but are now beneficial, have come to be associated quite strongly with the new mutant.

Once again the mechanism by which this has been brought about is natural selection working on an infinite gene-complex. Although industrial melanism in moths has provided evidence of a spectacular nature and has therefore been considered a phenomenon apart from normal evolutionary mechanisms, this should not be so. Exactly the same methods are used throughout the evolution of all living things. It is the *speed* at which it has happened that has impressed and misled us. This speed was dictated by the rate at which Man has, for the first time, changed the environment of parts of this Earth. It is indeed a comforting thought to know that, if in future he pollutes it in other ways, the mechanisms of life may be sufficiently elastic to meet the new situations.

Chapter Seven

Neodarwinism—the plain truth

Despite assertions to the contrary, Darwin's ideas about the origin of species hold good today

BRIAN CHARLESWORTH

NEODARWINISM—the Modern Synthesis—is the view of evolutionary mechanisms that emerged in the 1920s, 1930s and 1940s from the integration of Mendelian genetics into evolutionary biology. It is based on the finding of classical genetics that nearly all heritable variation is under the control of Mendelian genes located on the chromosomes. Because evolutionary change is a transformation of the variation present within a population at one time into differences between ancestral and descendant populations, the rules of inheritance place strong constraints on the processes of evolution. Those processes must involve the rather limited set of forces that can cause changes in the frequencies of different genes or gene combinations in natural populations. The most important of these are *mutation* (the spontaneous origin of new genetic variants, due to changes in the structure of the genetic material), *natural selection* (the differential survivorship or fertility of different variants), *gene flow* (the introduction of new genes into a population by immigration of breeding individuals), and *genetic drift* (random changes in the genetic composition of a population of limited size, the result of accidents of sampling during the passage of genes between successive generations). A notable omission from this list is the inheritance of acquired characteristics, which was long believed by evolutionists (including Darwin) to be an important agent of evolutionary change. We now know that, except in rare and special cases, the genetic material is so stable that characteristics acquired during an organism's lifetime cannot be transmitted to its offspring (see chapter nineteen).

Museum für Naturekunde

The Berlin specimen of Archaeopteryx *is clearly an intermediate between reptiles and birds. Do species evolve through a fairly complete series of intermediate stages or in evolutionary jerks?*

Unless past evolutionary change involved factors that cannot be studied experimentally at the present day, and which are therefore not open to scientific analysis, all evolutionary phenomena must be caused by the operation of one or more of the above forces. Hence, the differences we see, whether between local populations or geographical races of the same species, between related species, between genera, families and other higher taxonomic groups, must all result from causes of this sort. Because we cannot breed crosses between members of different higher taxonomic groups or between members of fossil populations and their descendants, and can only rarely cross different species, it is a matter of inference that differences between groups are due to the same set of processes that we can study at the within-species level. Systematists and palaeontologists have attempted to test this inference by asking whether evolution above the species level (*macroevolution*) exhibits features that are inconsistent with the predictions of genetic theory. The failure to find such inconsistent features is an important component of the Modern Synthesis.

The forces involved in macroevolution need not necessarily be identical with those involved in changes within a species (*microevolution*). It could, for example, be natural selection that is important in microevolution but species differences might originate by genetic drift. There have, however, been no particular reasons to expect such differences between different levels of evolution, and the prevailing view has been that differences between species and higher taxonomic groups are due to the cumulative effects of microevolutionary change. Most evolutionary biologists have accepted this view because it has a satisfactory theoretical basis, and agrees with a large body of data. Recently, however, it has been challenged by advocates of the theory of evolution by punctuated equilibrium, described in chapter eight. Gould argues that the fossil record shows that most lineages do not change much for thousands or millions of years (*stasis*), and that significant change is concentrated into geologically very short periods (*punctuational events*). Rapid evolution in small populations, involving both the acquisition of the inability to interbreed with the rest of the species (*speciation*) and morphological change, accounts for this pattern. Such changes, Gould argues, often involve mutations that have a relatively large effect on morphology, and they supposedly help to overcome constraints imposed by the homeostatic mechanisms of development, which restrict the potential for evolutionary change.

Many evolutionary biologists, myself included, think that these ideas are not sufficiently well supported either theoretically or empirically to warrant abandoning the view that microevolution is the stuff of macroevolution. We do not think that the formation of new species is a special class of evolutionary event. I shall now attempt to give some reasons for this position. Let us first consider the question of whether conventional microevolutionary processes seem adequate to explain the punctuational pattern of stasis alternating with rapid change; it is clear that this is at least sometimes observed in well documented fossil lineages, although palaeontologists disagree rather strongly about how typical it is compared with more evenly paced evolutionary change.

In the first place, it has been accepted since Darwin that evolutionary rates vary widely between different lineages and within the same lineage at different times. Punctuated equilibrium is simply an extreme version of a phenomenon that has repeatedly been discussed within the framework of neodarwinism. Nevertheless, it is legitimate to ask if observed patterns of evolutionary change are consistent with conventional theory; for instance, whether the rates of change we see in cases of rapid evolution are too high to be produced by the usual forces of microevolution. A difficulty here is that the

Rida Photo Library

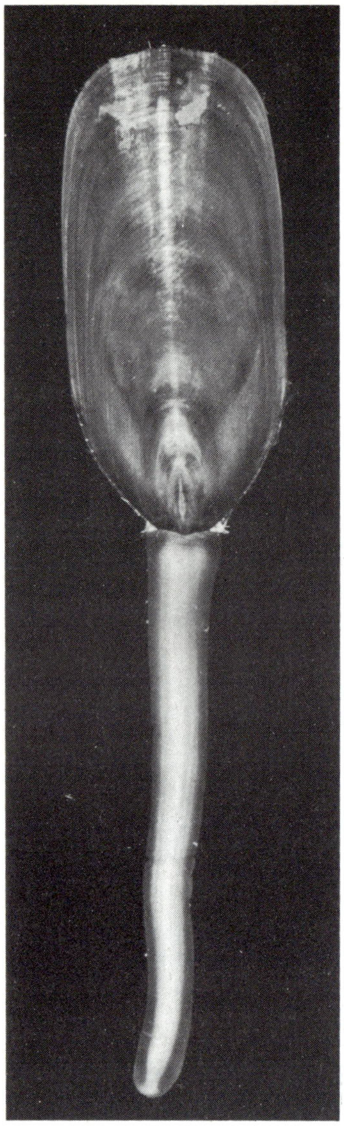

Lingula, *a lampshell, almost unchanged in 400 million years. The living animal (right) is very like the fossil (above)*

resolution of the fossil record is rather coarse compared with the time scale of microevolution. An average of 50 000 years between successive samples followed up a column of rock would be regarded as good in most studies of fossil populations, whereas in moths such as the peppered moth *Biston betularia*, dark forms (which are better camouflaged against predators in industrial areas) replaced light-coloured ones in less than 100 years (see chapter six). If we notice a significant difference between samples separated by 50 000 years, we obviously have no idea whether it evolved over 10 or 10 000 years. We can, however, ask whether the degree of divergence between successive samples is such that it could be produced by selection or genetic drift acting on variability present in the earlier member of each pair; if not we might need to appeal to new mutations of large effect as suggested in the punctuational model.

Studies of variability in fossil populations generally show that the averages of successive populations lie well within the range of variability of the earlier member of each pair. This is true even of Peter Williamson's Lake Turkana molluscs, which exemplify the punctuational pattern of evolution rather well. Human evolution, commonly regarded as one of the fastest evolutionary transitions on record, provides another example (see chapter three). *Homo erectus* and *H. sapiens* have mean cranial capacities (a measure of brain size) of about 920 ml and 1400 ml respectively, and almost certainly represent ancestral and descendant populations. One can calculate, from data on the variability of cranial capacity in fossil hominids, that about 1 in 500 individuals in an *H. erectus* population would have a cranial capacity of 1400 ml or more. Characters of this sort usually have a strong heritable component to their variation, which makes it easy to use standard genetic theory to show that selection for increased brain size could produce this magnitude of difference over a geologically very short period. Suppose, for example, that the difference in brain size between *H. erectus* and *H. sapiens*

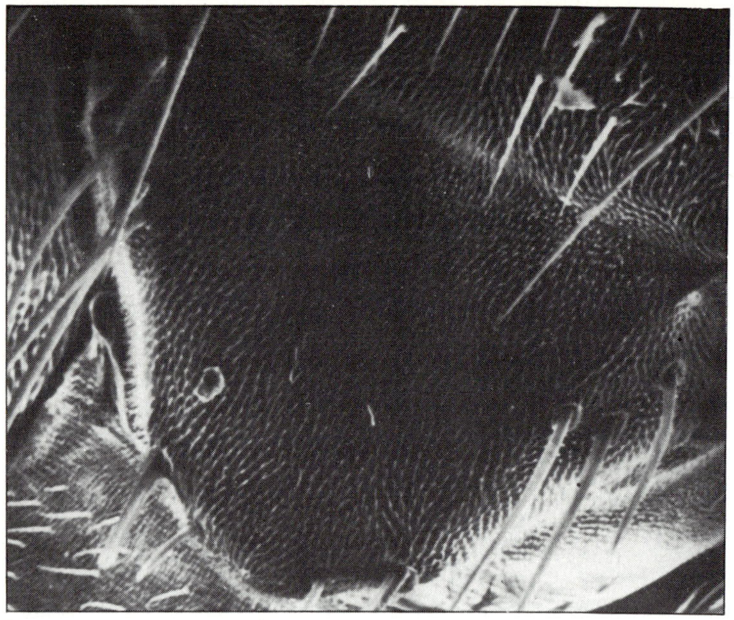

Figure 1 *Selection for extremes. The fruitfly above has 10 bristles on its scutellum, instead of the usual 4 (below)*

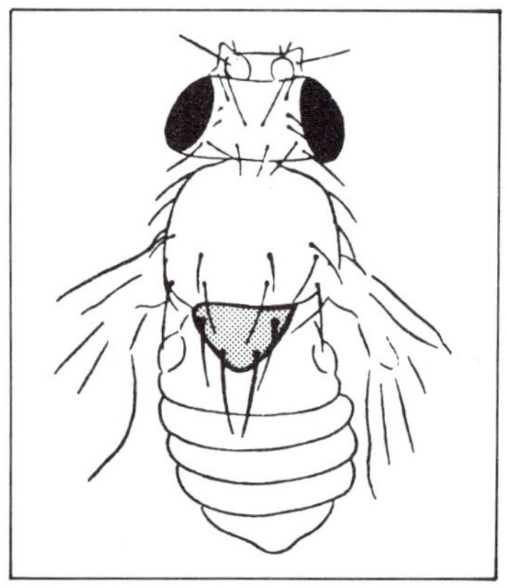

Rida Photo Library

evolved in only 20 000 years (1000 generations). A plausible model suggests that selection for increased brain size would eliminate only 1 in 500 individuals in each generation. This selection is extremely weak compared with that commonly enjoyed in artificial selection experiments in the laboratory, which have produced much larger changes in a variety of characters in 50 generations or less.

These considerations lead naturally to the question of why, if evolutionary change is so readily brought about, do there often appear to be long periods of stasis when little or nothing happens, at least at the level of characteristics that can be studied in fossil material. The extreme examples of this are organisms like the lampshell *Lingula*, which has changed little for the past 400 million years. There are three main possibilities: the species in question possesses little or no genetic variability in the characteristics concerned, so that evolutionary change is impossible; genetic variability exists, but is not subject to selection, and with the population so numerous that genetic drift is ineffective; finally, natural selection acts to preserve the status quo rather than to promote change.

The first possibility is implied in the explanation favoured by critics of neodarwinism, who invoke constraints imposed by the homeostatic mechanisms of development. Evolutionary stasis in a character would be inevitable if its development were so tightly controlled that genetic changes could not affect its expression. There are indeed many characteristics that show little variation under normal circumstances. Nevertheless, careful studies of sufficiently large population samples usually turn up variant individuals, and one can practise selection successfully. The bristles on the scutellum of *Drosophila* are a classic example of this; four large bristles are characteristic of the group and variants are extremely rare in natural populations. Nevertheless, it is possible by artificial selection to produce flies with up to 11 bristles (Figure 1). Constraints imposed by developmental mechanisms therefore seem unlikely to explain evolutionary stasis.

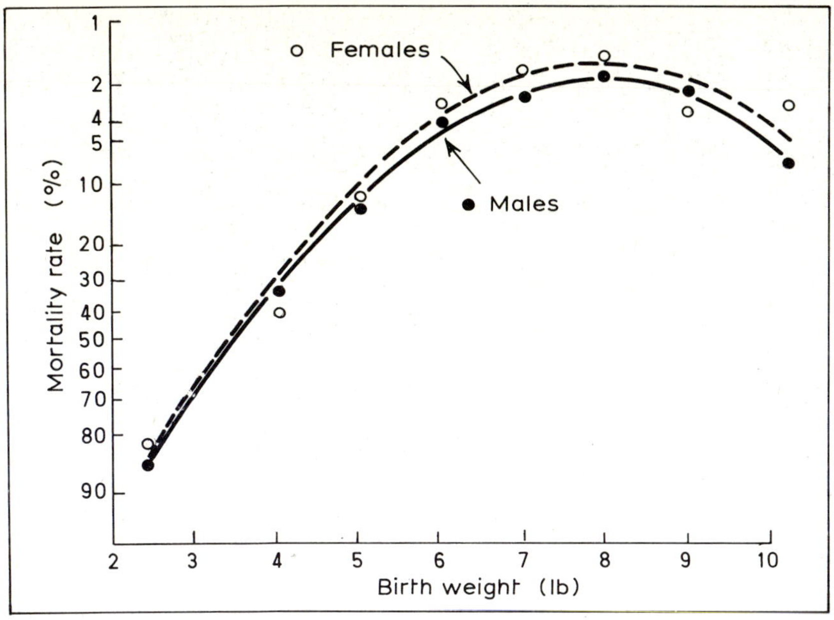

Figure 2 *Stabilising selection. Babies that are too heavy (right) and those that are too light (left) do not survive as well as babies of average weight*

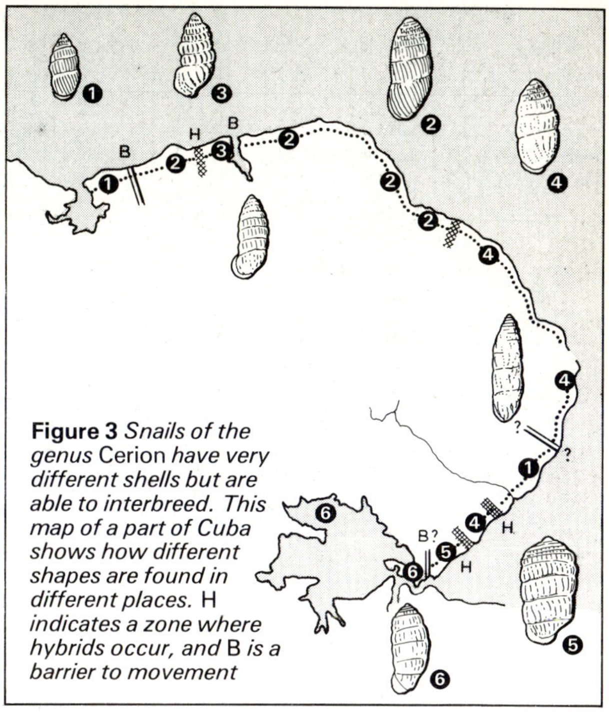

Figure 3 *Snails of the genus* Cerion *have very different shells but are able to interbreed. This map of a part of Cuba shows how different shapes are found in different places.* H *indicates a zone where hybrids occur, and* B *is a barrier to movement*

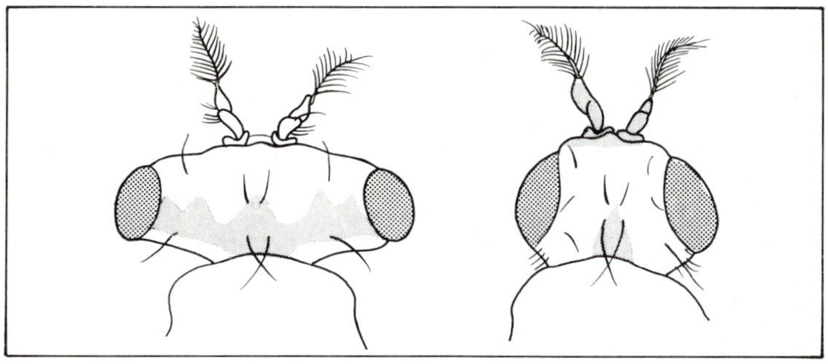

Figure 4 *Two very closely related* Drosophila *species—with no visible differences between their chromosomes—have very different eyes. The stalked eyes of* Drosophila heteroneura *(left) are the product of about 10 genes*

This does not mean that organisms are free to vary and evolve in any way they please, because the range of possible effects of genes on a character is limited by the existing developmental system, produced by millennia of evolution. Thus, it is clearly impossible to breed a race of six-legged horses, but breeders have had no difficulty in producing the wide diversity of types of limbs that characterise modern breeds of horse. Constraints imposed by the organisation of developmental processes are far more likely to be important in restricting the action of selection to relatively small changes in what has already been built up by previous evolution, than in triggering the sudden large changes in morphology suggested by punctuationists.

The most likely explanation of stasis is that the character in question is sufficiently well adjusted to its functional requirements that *stabilising selection* is operating; individuals with values close to the population mean have the highest fitness, whereas selection acts against more extreme individuals. Figure 2 shows a human example. Under stabilising selection, we would expect a population to evolve rapidly to a state in which the optimum value for the character coincides with the mean, and then to remain constant unless the optimum changes. Characters subject to stabilising selection would evolve mechanisms that tighten their control of development, thereby reducing the frequency of extreme individuals. This has been achieved experimentally in some artificial selection experiments. Developmental homeostasis is thus itself a product of selection.

It is, of course, a surprising fact that some characteristics of some organisms should apparently have an optimal value that has remained nearly constant over very long periods of time. It is important to bear in mind, however, that we are usually so ignorant of the genetic structure of fossil populations and of the relations between environment, fitness and morphology that we cannot provide explanations for any particular historical pattern of evolution. We can only inquire, as above, as to whether the broad patterns of evolutionary change that the fossil record reveals seem to be consistent with our expectations. Probably the most general relevant prediction of the theory of natural selection is that episodes of rapid evolution should coincide with periods when the direction of selection is changing; this seems to be borne out at many different levels of evolution. Insecticide resistance evolves in populations exposed to a new insecticide. The molluscs of Lake Turkana changed when the level of the lake altered. The *Drosophila* of Hawaii evolved an array of diverse species as they colonised an archipelago with numerous vacant ecological niches. And modern mammals underwent their period of most rapid evolution and diversification after the dominant land reptiles of the Cretaceous era went extinct. But what about speciation?

Speciation is the process by which a population comes to be incapable of effective interbreeding with related populations, and hence acquires a potentially separate evolutionary fate. The neodarwinian view is that speciation is usually the consequence of the evolutionary divergence of populations separated from each other by geographical barriers to interbreeding. If such divergence continues for a sufficiently long time, it is likely (although not inevitable) that the gene combinations selected independently in the different populations will eventually be so different that a hybrid between individuals from the populations will not function efficiently, or that the mating behaviour of the parents will be so different that hybrids will not even be produced. Species formation should thus be a fairly prolonged process, with a gradation of intermediate stages ranging from only weak isolation to complete inability to form hybrids. Taxonomic studies have amply documented that this is the case; there are many examples of geographic races of the same species that have diverged to such an extent that they are on the borderline of constituting new species, or are even totally isolated reproductively from at least some other races. There is no strong correlation between speciation and morphological evolution; many groups of species are morphologically virtually indistinguishable but reproductively isolated from each other, and in many other cases geographically separate populations look very different but show no evidence of isolation in breeding tests (Figure 3).

These facts suggest strongly that the differences between species are not unbridgeable gaps, but have been built up by the cumulative effects of the forces of microevolution. This conclusion is further strengthened by the results of genetic crosses between species that are closely related enough to be capable of interbreeding in the laboratory. There is no indication whatsoever from these crosses that the genetic basis of species differences is of a different class from within species differences. Most often, such differences involve numerous genes with relatively small effects on the character, just as in the case of variation within species (Figure 4). In particular, although it is true that species differ in chromosomal structure, there are many well-studied cases in which closely related species do not differ in this way, yet are morphologically quite distinct. Conversely, the intensively studied chromosomal variation within *Drosophila* species has no major consequences for morphological evolution. The important causal role that punctuationalists give to chromosomal rearrangements in evolution has no basis in fact.

Punctuationists could still argue that the differences between groups at the level just discussed are relatively trivial, and that microevolution is incapable of generating differences on the grand scale, such as those that distinguish different classes—mammals and reptiles. In particular, how do major evolutionary novelties—for example the viviparity, chewing jaw mechanism and teeth, hair and so on that distinguish mammals—arise? Do we need to appeal to mutations of large effect that break the old mould and provide the basis for rapid evolution of new structures? In most instances, there is no fossil evidence for intermediates linking major groups, probably because of the low chance of recovering fossils of relatively rare newly-evolved forms (a mere five specimens of *Archaeopteryx*, the only known intermediate between reptiles and birds, have been discovered). Nevertheless, there are cases in which a fairly complete series of intermediate forms is available, such as the transition from mammal-like reptile to mammal. This displays a graded series of changes that involve dentition, the jaw articulation, the structure of the ear bones, locomotion, temperature regulation, and so on. Taxa such as *Eozostrodon* possess a blend of reptilian and mammalian characteristics. These evolutionary changes involve differences in the sizes and proportions of teeth and bones of precisely the kind I discussed for within- and between-species variation.

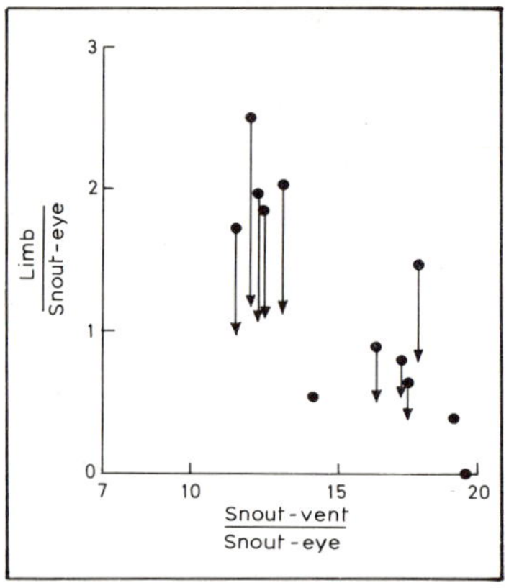

Figure 5 *The lizard genus* Brachymeles *shows a gradual reduction in the length of its limbs compared with its body size. Each line is one species, the circle the front limb and the arrowhead the back limb. The horizontal axis is a measure of the elongation of the body, an adaptation to burrowing; lizards that make burrows have shorter legs*

R. A. Fisher pointed out 50 years ago that a complex adapted structure involves the integration of numerous different components, so that a large change in any single part is overwhelmingly more likely to destroy its interrelationship with others than to lead to an improvement. The evolution of a new structure is thus far more likely to occur by successive small changes in individual components than by large mutational steps. The loss of limbs, which has occurred independently in several groups of lizards adapted to a burrowing mode of life, provides compelling evidence for this, and one can assemble graded series of related species that show progressive reduction in limb size (Figure 5). The evidence favours a gradual evolution towards limblessness, and this final example shows that we need no special mechanisms to explain the pattern of evolution. The processes of microevolution are more than adequate.

Chapter Eight

Punctuated equilibrium—a different way of seeing

Was Darwin wrong in his major premise, that the origin of species is simply the origin of adaptation on a larger scale? One of the founders of an alternative theory of how species arise thinks he was

STEPHEN JAY GOULD

PUNCTUATED equilibrium is a theory about the way that species arise in geological time. As such, it is about the tempo and mode of evolution. It holds, speaking of mode, that significant evolutionary change happens at the same time as events of branching speciation, and not primarily through the *in toto* transformation of lineages (classical anagenesis). The changes that distinguish one species from another arise when that species arises. It maintains, speaking of tempo, that the proper geological scaling makes speciation geologically instantaneous, and that, following this rapid origin, the morphology of most species fluctuates only mildly during a period of stasis that usually lasts for several million years. ("Geologically instantaneous" is a fuzzy term, so I make the operational suggestion that it be defined as one per cent or less of the species' later existence in stasis. This permits up to 100 000 years for the origin of a species with a subsequent life span of 10 million years, though I believe that most events of speciation occur much more rapidly.)

I am not saying that punctuated equilibrium is the *only* mode of speciation. As with all major issues in natural history, support for punctuated equilibrium relies on an argument about relative frequency, not a claim for exclusivity. Gradual, phyletic transformation can and does occur. I claim merely that its relative frequency is low and that punctuated equilibrium is the predominant mode and tempo of evolutionary change.

It is hard to put a number on "low" and "predominant". Would an off-the-cuff claim for 90 per cent be sufficient to rescue me from a charge of winning my own argument by defining it for each victory? If God appeared and informed us that 48.647 per cent of significant evolutionary events had occurred via punctuated equilibrium, I would be more (but only slightly more) than half satisfied. To be half right, however, is a blessing in a complex world.

By my one per cent criterion, tens of thousands of years are usually available for the origin of full reproductive isolation

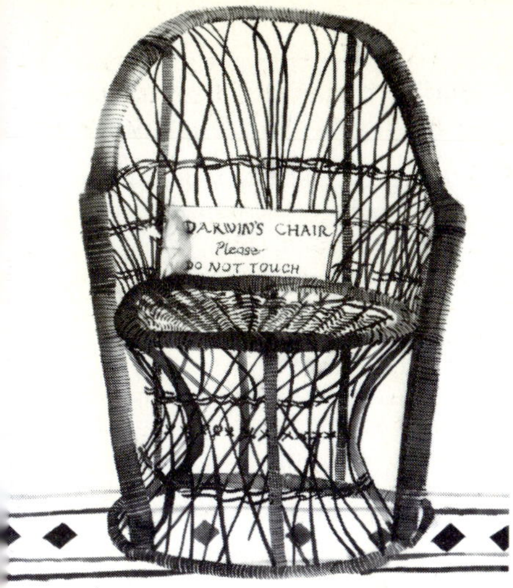

between new species. Because times of this order are sufficient for almost any model of speciation (except "dumb-bell" allopatry, where a population splits into two roughly equal parts and each diverges slowly thereafter), punctuated equilibrium *per se* does not suggest or specify any particular mode of speciation. In fact, Niles Eldredge and I wrote our first paper in 1972 to argue that the most orthodox form of Ernst Mayr's "peripheral isolate" model implies punctuated equilibrium in geological scaling.

Mayr's idea was that species arose when a small part of a larger population became isolated, by a river, a mountain, or some other geographical feature. This small peripheral isolate, as it was called, was able to evolve more rapidly than the large major population, and if conditions changed so as to favour the peripheral isolate, would be able to sweep through the home range and displace its parent species. (This is called allopatric speciation because the diverging species occupy different locations.) Eldredge and I simply suggested that if Mayr's model were true, we would not expect to see the intermediate forms of the evolving peripheral isolate captured in the fossil record.

In this light, I have been vexed that several colleagues have attacked punctuated equilibrium by casting doubt upon rapid parapatric and sympatric modes of speciation. I shall, of course, be pleased if these modes are in fact established at reasonable or high relative frequencies—the faster the better for punctuated equilibrium. But because punctuated equilibrium is consistent with the standard Darwinian model of the modern synthesis—indeed it was developed in this context (for these rapid modes either had not been proposed or had few supporters when we first wrote in 1972)—its vindication is not dependent upon the outcome of current debates about speciation. This does not mean, as other colleagues have charged, that punctuated equilibrium is soft, squishy or unfalsifiable because most common modes of speciation are consistent with it. It means only that punctuated equilibrium is a phenomenon of different scale, and that its support or falsification will be tied to events at geological

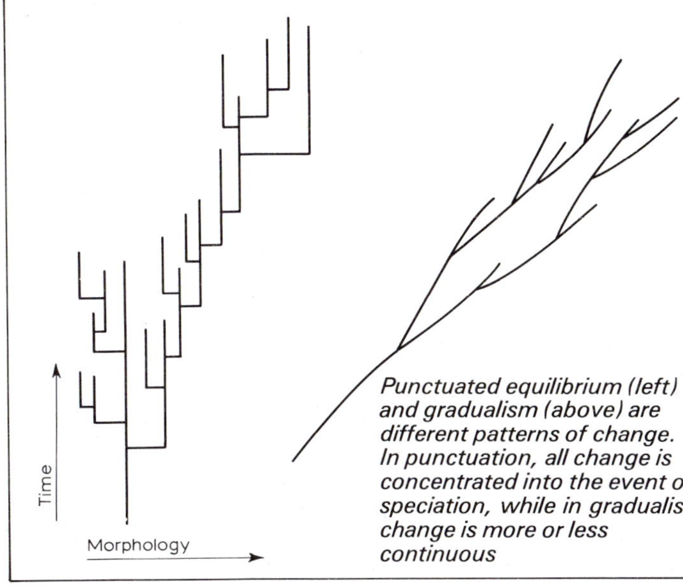

Punctuated equilibrium (left) and gradualism (above) are different patterns of change. In punctuation, all change is concentrated into the event of speciation, while in gradualism change is more or less continuous

A minute portion of Darwin's beetle collection

Drawings by Chris Corr

dimensions (including the relative frequency of stasis; the persistence of ancestors after punctuations; the efficacy of species selection in producing evolutionary trends).

Punctuated equilibrium is not a theory of macromutation; indeed, it is not a theory of any genetic process (although its geological geometry, if vindicated, exerts some constraint upon genetic modes. Gradual and sequential substitution of genes will not be a good model for the origin of higher taxa if stasis be prevalent.) It is a theory about larger-scale patterns: the geometry of speciation in geological time. As with ecologically rapid modes of speciation, punctuated equilibrium welcomes macromutation as a source for the initiation of species—the faster the better. But punctuated equilibrium clearly does not require or imply macromutation, for we formulated it as the expected geological consequence of straightforward Mayrian allopatry.

I do feel that certain forms of macromutational theory are legitimate, and I have supported them, though not in the context of punctuated equilibrium (I do have other interests, after all). I doubt that macromutation initiates species with a high relative frequency, but even rare occurrence may produce important evolutionary results because big biological rearrangements are themselves so uncommon. I regard forms of macromutation which include the sudden origin of new species with all their multifarious adaptations intact *ab initio*, as illegitimate. So is species origin by drastic and sudden reorganisation of the entire genome of an ancestral species. Legitimate forms include the rapid saltatory origin of a key feature (around which subsequent adaptations may be moulded) and large outward changes caused by small genetic changes that affect rates of development in early ontogeny with cascading effects thereafter.

Few evolutionists recognise (and his colourful and absolutistic phrases didn't help his case) that Richard Goldschmidt set his "hopeful monster" theory in this legitimate context. Goldschmidt was primarily interested in how development constrains and facilitates macroevolution. He defined hopeful monsters as phenotypic products of small genetic changes that have their impact early in ontogeny. Some small genetic change had a massive effect, producing a very new form, a monster. Cascading effects arise from potential alternative pathways of development that are already contained in inherited norms of reaction. (Some of these alternative pathways can be seen in teratologies, phenocopies, mutants and simple sexual differences.)

Goldschmidtian monsters may be hopeful because normal development does tend to channel perturbations along viable, if discontinuous, routes. (Most monsters, of course, are hopeless because they are not so regulated by development, and these monsters play no role in evolution.) Genetic differences between hopeful monsters and normal forms are minor, so breeding may not be impaired; under certain population structures (discrete small groups that inbreed), small populations of homozygous hopeful monsters may be established. The ultimate success of a hopeful monster will depend upon the subsequent development of stabilising adaptations around whichever key features have appeared in the sudden shift.

An example is Hugh Iltis's theory for the origin of the corn (maize) ear, presented at a meeting of the Society for the Study of Evolution in 1981. Iltis postulates a hopeful monster of pure (and legitimate) Goldschmidtian form; the ear evolves by making use of developmental pathways—in this case, ordinary sexual differences—already present in teosinte (the presumed ancestor of domesticated corn). Iltis sees the central spike of the male corn tassel as the homologue of the female teosinte "ear". He regards the modern corn ear, which is female, not as the gradually enlarged teosinte ear, also female, but as the saltationally transformed male central spike. Teosinte female ears are carried laterally to the lateral branches off the stalk, while the modern corn ear is at the tip of the lateral branch. The male spike of teosinte, like the female corn ear, is at the end of the lateral branch, and Iltis argues that if the lateral branch shortened it would bring the male tassel into a zone of feminisation where, because it is at the branch tip, it might enlarge itself suddenly and suppress the growth of female ears below. The whole process might be accomplished by no more new genetics than would be needed to shorten the lateral branch—given the "old" genetics of sexuality and development. Iltis even suggests that no internal genetic change at all would be needed to shorten the branch—various viruses and smuts may produce this effect. (Some teratologies of modern corn can produce a female ear in place of the male central spike.) Human selection would preserve, propagate and "improve" such a teosinte hopeful monster.

Teosinte grass

Chicago Natural History Museum

I cannot judge Iltis's case, but as a general argument it is certainly legitimate in theory, interesting and worthy of pursuit. And just as certainly, our subtle gradualistic biases have impeded the formulation of such alternative propositions by directing attention to the gradual enlargement of teosinte ears.

These legitimate styles of macromutation are related to punctuated equilibrium only in so far as both represent different and unconnected examples of a general way of thinking that I have called punctuational (as opposed to gradualist or continuationist thought). I take it that no one would deny the constraining impact that gradualistic biases have upon evolutionary theorising. Punctuational thinking focuses upon the stability of structure, the difficulty of its transformation, and the idea of change as a rapid transition between stable states. Evolutionists are now discussing punctuational theories at several levels: for morphological shifts (legitimate macromutation), speciation (various theories that offer rapid reproductive isolation) and general morphological pattern in geological time (punctuated equilibrium). These are not logically interrelated, but are manifestations of a style of thought that I regard as promising and, at least, expansive in its challenge to conventional ideas. Any manifestation of the various punctuational theories may be true or false, or happen seldom or often, without affecting the prospects of any other. I do commend the general style of thought as a fruitful source for hypotheses.

If the geometry of punctuated equilibrium prevails within the history of most clades (groups descended from a single ancestor), then traditional arguments for evolutionary trends, the most important phenomenon that palaeontologists study, must fail. In the traditional argument, trends are the products of anagenetic transformation with lineages, brought about by natural selection. That is, the whole lineage participates in the trend. Of course, no one has ever held that sustained trends occur in the complete absence of any speciation within a clade. The standard phyletic model does not deny speciation, but simply grants it no role in *directing* the trend. In phyletic models, speciation entails no alteration in the direction of change nor any acceleration of rate, and so it does not enhance the directional component of the trend. Its role, instead, is to iterate a favourable adaptation (which arose as a gradual phyletic change) into several lineages, thus aiding the adaptation's spread and compensating for any extinction. (After a split, ordinary phyletic evolution can occur in two lineages rather than just one.) This picture lies behind the traditional separation of cladogenesis (the pattern of branching) from anagenesis (the pattern of change within a branch)

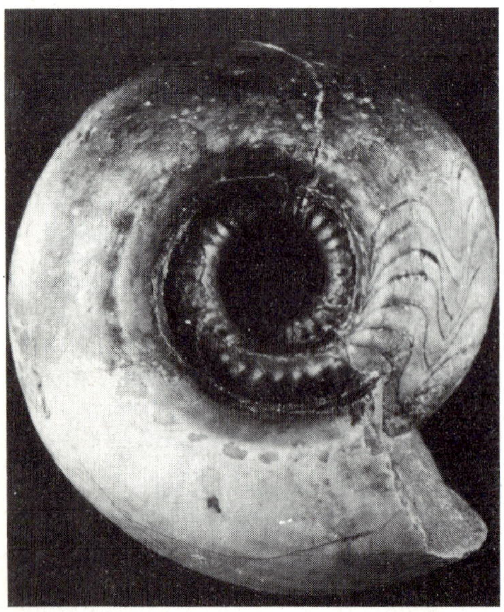

The suture lines joining the chambers of ammonites show a progressive change in complexity as the ammonites evolve,

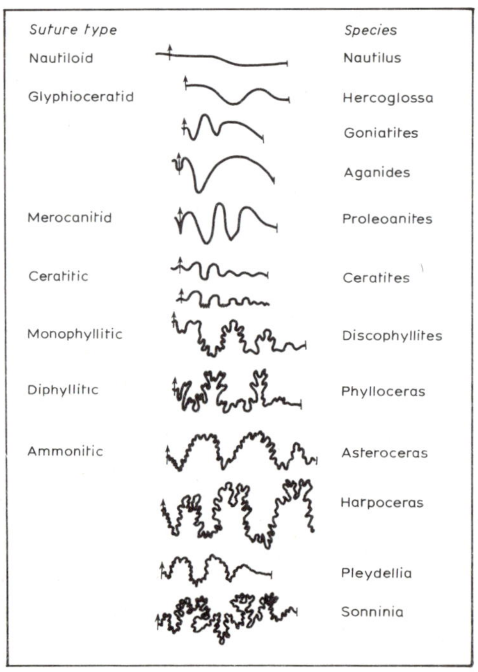

from the very simple pattern at the top to the more convoluted one below. Why? Is there any adaptive answer?

Bruce Coleman/Oxford Scientific Films

and also supports their opposed significances as generators of "diversity" and "progress" respectively.

Two classical trends are the increasing complexity of suture lines between chambers in ammonite shells and increasing symmetry of the cup in Palaeozoic crinoids. If we explore the speculative explanations offered, we find that these explanations presuppose a direct advantage for some particular morphology under a phyletic regime of natural selection. The complex sutures strengthen the ammonite shell, for example; or perfect radial symmetry that permits sessile organisms like crinoids to collect food from all directions. We also find that such explanations have been notoriously unsuccessful, a plethora of contradictory speculations that never brings resolution. Perhaps the problem lies deeper than our failure to devise a good adaptive story; perhaps we must challenge the basic (and usually both unstated and unrecognised) assumption that proper explanation must lie with such a story, if only we could find the right one.

Under the geometry of punctuated equilibrium, there is no phyletic component to direct a trend; species arise (in geological time) with their differences established at the start, and do not change substantially thereafter. Trends must therefore be the product of a higher order process that operates upon the differential birth and death of species, which must now be considered as entities. In other words, species play the same role in an evolutionary trend that individual organisms, which do not change evolutionarily during their life, play in the adaptive processes of microevolution. Species selection, as this process is called, acts by differential origin and extinction, to direct evolutionary trends within clades (macroevolution), just as natural selection, acting by differential birth and death of individuals, directs evolutionary change within populations (microevolution).

The geometry of punctuated equilibrium means that we cannot extrapolate natural selection within populations to produce evolutionary trends because trends are a product of the differential fate of species considered as stable entities (that is, as true "individuals" at their higher level). Under phyletic models, such extrapolation can be made (as the cited speculations for ammonite sutures and crinoid symmetry demonstrate), and no separate theory of macroevolution is required (see chapter seven). Macroevolution under punctuated equilibrium is, as Steven Stanley argues, decoupled from microevolution—by which claim neither he nor Eldredge and I ever intended to advance the despairing hypothesis that macroevolution is different in kind from microevolution, or that microevolution, in principle, can illuminate nothing about macroevolution. The failure of pure extrapolation betokens only the need to consider other forces and information, not the irrelevance of the lower level. Uncoupled railway cars still run along the same track and exert constraints upon each other's speed.

I believe that species selection has two important implications for evolutionary theory. First, it expands markedly the classes of hypotheses available for explaining evolutionary trends. Previously, an unexamined assumption of phyletic transformation required that one view morphological trends as the direct product of natural selection acting upon the advantages of individual organisms (stronger shells, better access to food, etc.). But if trends record instead the differential success of species, then we have to consider a whole new set of previously unexplored explanations.

Trends may occur simply because some kinds of species speciate more often than others, not because the morphologies so produced have any advantages under natural selection (indeed, such a trend will occur even if extinction is completely random). Or some species may survive longer than others because they inhabit a certain environment, not because their morphologies are "better" in any conventional sense.

Punctuated equilibrium, I have argued, implies that evolutionary trends cannot be reduced to natural selection acting upon individual organisms; a higher level of explanation must be acknowledged as legitimately independent in evolutionary theory. Explanations at this level, because they are so heavily dependent upon species selection, have analogues at the lower level of natural selection, but species selection is not the same thing as natural selection; for true group-level effects often exist and the analogies yield interesting differences.

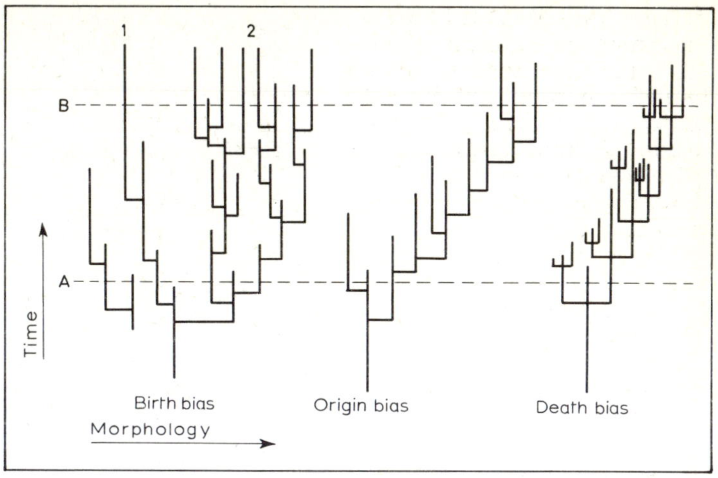

Species selection. Birth bias implies that a particular morphology splits more often, death bias that it goes extinct more quickly. Origin bias means that species tend to split in one direction rather than another

This hierarchy of levels is a contingent fact of the empirical world, not a mere issue of semantics or methodological styles. We can construct a possible and non-hierarchical world in our thoughts; indeed, evolutionary theory has operated for many years under the assumption that such a world exists. The extrapolationist, phyletic model requires no hierarchy, processes at each level being the same; if it reflected our world, reductionism would triumph. Punctuated equilibrium is the empirical claim that invalidates such a model, requires hierarchy, and questions reductionism. This is its major importance in the structure of evolutionary theory.

The issue is larger than the independence of macroevolution. It is not just macroevolution *vs* microevolution, but the question of whether evolutionary theory itself must be reformulated as a hierarchical structure with several levels—of which macroevolution is but one—bound together by extensive feedback to be sure, but each with a legitimate independence. Genes, bodies, demes, species and clades are all legitimate individuals in some situation, and our linguistic habit of equating individuals with bodies is a convention only. Each type of individual can be a unit of selection in its own right. Natural selection operating on bodies will not encompass all of evolution. Genes are units of selection in the hypothesis of "selfish DNA", demes (breeding groups) are units in Sewall Wright's shifting balance theory. Species selection is one new level among many, and I believe that evolutionary theory needs this expansion.

Natural selection operates either by selective death or selective birth. All evolutionists appreciate this of course, but a cultural legacy that long antedates Darwin has led us to emphasise death and the literal struggle for existence; nature red in tooth and claw, in Tennyson's phrase. We view evolutionary change by differential birth as an odd process, occurring here and there to be sure, but not contributing to "progress" or increasing complexity.

Both Eldredge and I and Steven Stanley followed this restrictive legacy in our original formulations of species selection. We acknowledged the analogue of differential birth (speciation) in our words, but formulated our discussions almost entirely in terms of differential death (extinction). The oft-repeated statement that speciation may be random

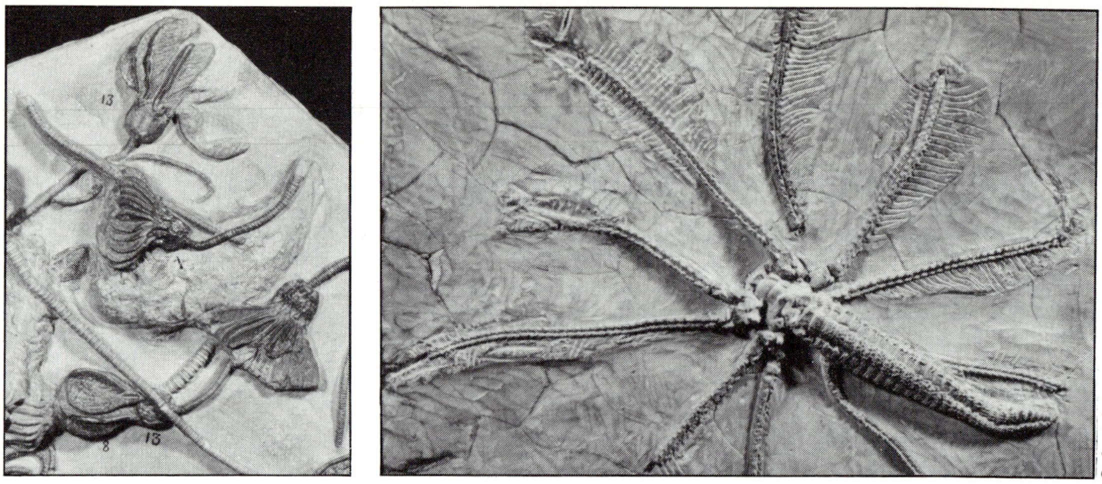

Oxford Scientific Films

Fossil crinoids. They become more symmetrical as they evolve. Like the ammonite suture, there is no convincing reason to explain why this should have happened

with respect to the direction of an evolutionary trend reflects the tight link that we made between species selection and differential extinction. I now believe that this formulation was inadequate in its misplaced emphasis, and that evolutionary trends powered by differential birth are both common and more interesting in their unconventional implications.

A trend can occur via differential speciation in two ways, only one of which represents species selection. In "birth bias", a clade contains two or more kinds of species (analogues of gene alleles) at an initial time *A* and the increased representation of one kind at a later time *B* arises only from its higher rate of speciation. This is species selection by differential origin. In "origin bias" speciation is more likely to occur in one direction than in others. This is not species selection, but represents the higher-level analogue of mutation pressure, or differential direction of variation itself. Speciation in macroevolution is the analogue of birth in microevolution. Therefore, if new species arise with morphologies that are predictably different from their parents, an evolutionary trend can be powered by a bias in the raw material itself, an analogue to mutation pressure in microevolution. But in Darwinian explanations the randomness of the raw material is a primary ingredient. In short, the two processes of differential origin produce non-Darwinian effects. Birth bias utilises species as units of selection, and origin bias establishes a bias in directional production of raw material, the analogue of mutation pressure, as an important feature.

These cases of differential origin also illustrate important differences in style and emphasis between micro- and macroevolution. In microevolution, so many individuals (bodies) are produced with so little average difference among them, that both differential birth of bodies with a favoured allele and differential conversion to that allele by mutation pressure can easily be swamped by natural selection acting via the death of bodies. In this sense, our conventional feeling about the power of natural selection is contingent upon the existence of copious and similar individuals within a bounded population. In macroevolution, we have far fewer individuals (species) within our bounded unit (clade) and the average differences among individuals are far greater. In this situation, selection upon death (extinction) may not be able to overcome either a birth or an origin bias, and much of the "creativity" of macroevolution may lie in the generation of variation—that is speciation—itself, not in the differential culling of this variation.

The differences and similarities between macro- and microevolution have been a contentious issue within evolutionary theory from the very start. The absolute nature of such a separation lies at the basis of our first coherent system of evolution, that of Jean Baptiste de Lamarck. Lamarck contrasted progress up the ladder of complexity, mediated by an internal "force that tends incessantly to complicate organisation", with tangential adaptations induced by the external "influence of circumstances", transferred to heredity by the inheritance of acquired characters. Speciation and adaptation were distinct processes. This tradition of an absolute separation persisted throughout the 19th century in such macroevolutionary theories as orthogenesis and racial life cycles, in which the fact that the end results are foreordained renders natural selection of the moment irrelevant to any larger trend. Goldschmidt revived this tradition in our century when he argued for a complete separation between geographic variation and speciation, and for a different genetic style behind the origin of species and higher taxa.

Perhaps Darwin's greatest achievement lay in counteracting this tradition (he did not after all invent evolution itself). Darwin provided a strong argument for continuity and the reduction of macroevolution to natural selection acting upon individual bodies, an argument which provided a powerful model. For the first time, all the data that we can see and manipulate because they operate at our time scale (artificial selection, allelic change in natural populations) were no longer epiphenomena to a grand and mysterious cosmic directionality, but made up the very stuff of evolution itself.

How can we mediate between these two positions? I do not wish to advocate the despairing conclusion that micro- and macroevolution are absolutely separate in principle and that nothing about one illuminates the other (the tradition espoused by Lamarck and Goldschmidt). But shall we accept Darwinian continuationism at the price of ignoring the legitimately independent properties of levels higher than natural selection operating upon organisms? Must the choice be either despair and acknowledgment or ignorance and bliss?

I believe that the concept of hierarchy resolves this pseudoproblem by incorporating the favourable features of both positions. Levels do not achieve their independence because some fundamentally new genetic process emerges at their scale. Rather, the same basic processes of variation and selection operate throughout the hierarchy. But they work differently upon the varying materials (individuals) of ascending levels in a discontinuous hierarchy. We cannot learn everything we need to know about evolutionary trends by studying what happens within demes, if only because species can act as units of selection. Important ties of feedback unite all levels, but the new modes emerge at higher levels and reduction to natural selection upon organisms will not render all of evolution. Nothing about microevolutionary population genetics, or any other aspect of microevolutionary theory, is wrong or inadequate at its level. Little of it is irrelevant to students of macroevolution. But it is not everything.

When a proper hierarchical theory is fully elaborated, it will not be entirely Darwinian in the strict sense of reduction of natural selection acting upon organisms. Yet I suspect that it will embody the essence of Darwinian argument in a more abstract and general form. We will have a series of levels with a source for the generation of variation and a mode (or set of modes) for selection among individuals at each level.

The superseding of strict Darwinism may establish the Darwinian style of argument in its most general form as the foundation for a truly synthetic theory of evolution.

Chapter Nine

The reptiles that became mammals

Biologists traditionally have considered that the fossil record is too incomplete to contribute in detail to theories of evolution. But the fossil record of the mammal-like reptiles is certainly convincing —and if taken at its face value it gives rise to some quite new evolutionary concepts

TOM KEMP

IDEAS about the ways in which evolutionary changes take place have come almost completely from studies of living organisms—their genetics, ecology, distribution and so on. Fossils have played only a limited part in formulating evolutionary theory because the biologist can never be certain that all the critical stages are preserved in any particular part of the fossil record. So biologists have tended either to fit fossil evidence into theories derived from knowledge of living organisms or else—and these form the majority—to assume that the fossil record is too incomplete to be helpful.

Lately, this trend has been reversed. Several palaeontologists have tried to assume that the fossil record gives a better view of evolving life than had been supposed, and have then looked at the way such fossil evidence might suggest modifications to existing views of evolutionary theory. For example, the fossil record does not usually show a gradual transition from one form to another. Rather, a species remains the same for a long period of time and then, quite abruptly, is replaced by a different, but closely related, species. Rather than blaming the absence of transitional fossils on "gaps" in the fossil record a number of biologists categorised as "punctuationists" argue that evolutionary change did indeed occur, as the fossil record suggests, in a series of jumps. Most evolution is concentrated into brief bursts as new species arise from small isolated populations of other species; the intermediates stand little chance of being captured by the vagaries of fossilisation. In this chapter, I shall examine the fossil record of one particular group of animals, the mammal-like reptiles, on the basis that the record does provide a true picture, to see if these fossils can provide clues about how evolution takes place.

The mammal-like reptiles, or Synapsida, from which the mammals eventually evolved, are particularly amenable to this sort of endeavour. The first stage is to survey their pattern of evolution. The Synapsida existed from the Middle Carboniferous, some 315 million years ago, through to the end of the Triassic, 195 million years ago, at which time the first mammals had just appeared. The fossil record of the mammal-like reptiles is remarkably complete throughout this span of some 120 million years, with no very large gaps. What is more, the group shows a wide and complex adaptive radiation—that is, they evolved from a single stock lineage into many different types filling many different ecological niches. Overall, the Synapsida became, and for a long time remained, the dominant land animals.

The mammal-like reptiles evolved through three distinct phases of adaptive radiation. The first of these consisted of the Pelycosaurs, and the earliest forms are from Middle Carboniferous deposits in Nova Scotia. *Archaeothyris*, an early pelycosaur, was a lizard-like animal about 50 cm long that differed very little from the primitive reptiles alive at the same time; the major distinction is a small space—the temporal fenestra—in the cheek region of the skull. This little space is important because it permits the muscles of the jaw and the jaw itself to become more efficient. Indeed, the temporal fenestra is a hallmark of all the mammal-like reptiles which persists right through to the mammals; so all the synapsids must have evolved from an ancestral form not unlike *Archaeothyris*.

Pelycosaurs diverged into a number of groups, each of which reflects a different type of adaptation. The caseids and edaphosaurids were plant eaters, with blunt teeth and short, stout jaws. They were the first successful group of land-dwelling, four-footed herbivores. The ophiacodonts included fish-eaters, with teeth somewhat like those of a crocodile. And the sphenacodonts were the major carnivorous reptiles of the time, with powerfully built skulls and prominent canine teeth. Some of them were pretty large, like the "finback" *Dimetrodon,* 3-4 metres long.

Around 260 million years ago, at the end of the Early Permian, the pelycosaurs suddenly became extinct, except for a few caseids and sphenacodonts which just held on into the Late Permian before they too disappeared. At the same time, the start of the Late Permian, the first members of the next phase of the synapsid radiation appeared. These were the therapsids, which evolved from an unknown lineage of the sphenacodonts.

All the therapsids were distinctly more mammal-like than the pelycosaurs. They had larger temporal fenestrae, which indicates more powerful jaw muscles, and a more advanced form of locomotion. A specialised jaw is one feature that sets mammals apart; the way they move is another. The early reptiles, like the lizards and crocodiles today, generally rested their bellies on the ground. When they walked, it was with their limbs sprawling out to the side; this places a severe strain on the legs which have to hold the body up. Mammals brought the legs under the body, which reduces the strain on them and lifts the body free from the ground so that movement becomes much more efficient. The therapsids clearly represent a trend towards this mammal-like form. Even at their first appearance, the therapsids had already diverged into several groups, and during the course of the late Permian they came to occupy a set of ecological niches very similar to those previously occupied by the pelycosaurs.

The dinocephalian therapsids were rather primitive large

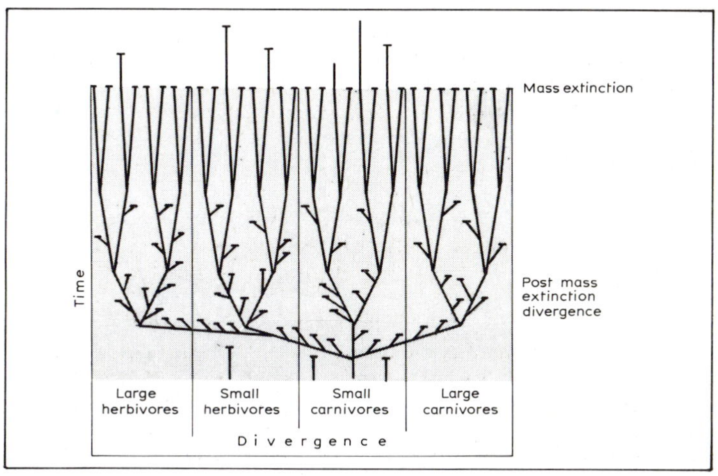

A typical adaptive radiation. The lines are evolving lineages of the main ecotypes. Immediately after a mass extinction, the rate of species turnover is high, and some kinds of lineages survive better than others

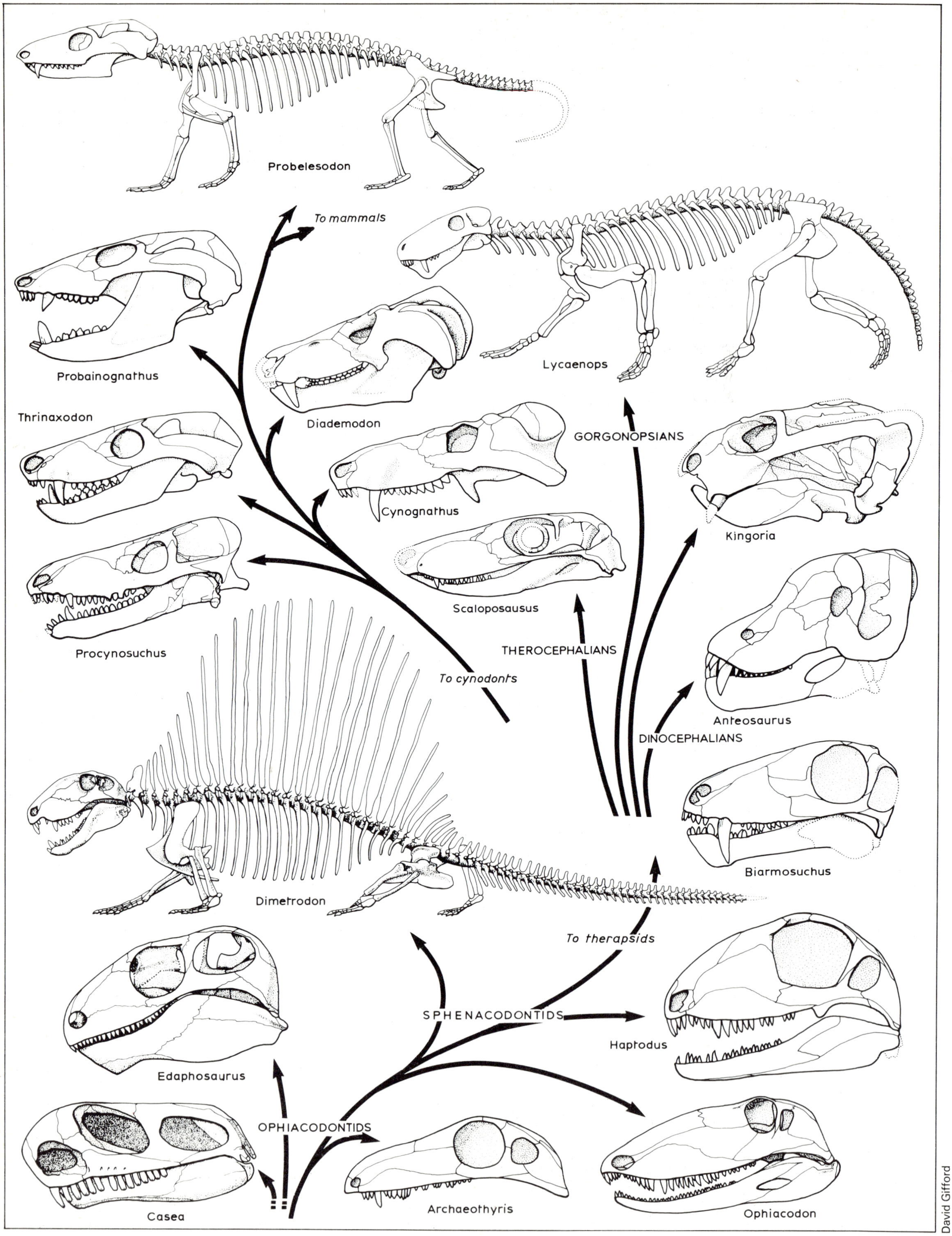
Probelesodon
To mammals
Lycaenops
Probainognathus
Diademodon
Thrinaxodon
GORGONOPSIANS
Cynognathus
Kingoria
Scaloposausus
Procynosuchus
THEROCEPHALIANS
To cynodonts
Anteosaurus
DINOCEPHALIANS
Biarmosuchus
Dimetrodon
To therapsids
SPHENACODONTIDS
Edaphosaurus
Haptodus
OPHIACODONTIDS
Casea
Archaeothyris
Ophiacodon
David Gifford

herbivores and carnivores, and they actually became extinct during the Late Permian. The gorgonopsians were the main carnivorous group; they became quite large, and had huge canine teeth. The main herbivores were the enormously successful dicynodonts, with many species, large and small. The dicynodonts lost most of their teeth, which were replaced by a horny, turtle-like beak, although many species did retain a pair of large upper tusks. There were also the therocephalians, a more mixed collection of species, mostly carnivorous, which included a number of very small, presumably insectivorous, types.

The final group of therapsids to appear in the Late Permian is made up of uncommon, and quite small, animals like *Procynosuchus*. These are the cynodonts, and they form the next link in the synapsid chain, because at the end of the Permian, 230 million years ago, most of the mammal-like reptiles abruptly became extinct. The only survivors were a few lineages of dicynodonts and therocephalians—and, most importantly, the cynodonts.

Cynodonts were even more like mammals than were the previous therapsids. They carried on the theme of jaw improvements, developing complex multi-cusped teeth and more powerful musculature. From this Early Triassic stock of cynodonts arose the third phase of adaptive radiation of the mammal-like reptiles. Again, the basic stock diverged into a variety of types; large and small herbivores and carnivores, and as the Triassic went on the fossils come to look increasingly mammalian. But the cynodonts too became extinct at the close of the Triassic, leaving only their tiny descendants—true mammals—to persist into the Jurassic and beyond.

Each of the three phases of synapsid evolution—pelycosaur, therapsid and cynodont—shows a basically similar evolutionary pattern, which suggests that fundamentally similar underlying causes determined the progress of evolution in each case. By considering only those characteristics of the pattern that are common to all three phases, we can come up with a generalised, or "model", adaptive radiation.

This general model incorporates the following eight features or rules:

Rule 1. The radiation begins after most, but not all, the lineages of the previous radiation have become extinct. This is the phenomenon of mass extinction.

Rule 2. The radiation begins from one of the surviving lineages of small carnivores.

Rule 3. The main groups of the new radiation are more or less fully evolved when they first appear in the fossil record, and they change relatively little in the course of the radiation.

Rule 4. All members of the new radiation are more advanced (that is, more mammal-like) than any members of the previous radiation.

Rule 5. Each radiation produces essentially similar kinds of animals each adapted to complementary ways of life: the ecotypes include large and small herbivores and carnivores.

Rule 6. Within each lineage, species keep going extinct abruptly, only to be replaced by new, similar species. This is the phenomenon of species turnover.

Rule 7. Minor mass-extinctions of several unrelated lineages occur from time to time during a radiation.

Rule 8. The end of the radiation is marked by the simultaneous extinction of all the species and most of the lineages composing it. A few lineages, mainly of small forms, survive this mass extinction.

Those are the rules for mammal-like reptiles. The ultimate aim of the palaeontologist must be to interpret the causes that underly the observed rules of evolution. Admittedly this job has not progressed very far as yet, but we can go some way towards providing an interpretation.

There are three variables that must be involved in any such interpretation, namely 1) rates of speciation and extinction in a lineage, 2) rate of morphological change within a lineage, and 3) direction of morphological change within a lineage.

Speciation is obviously important. Rule 6 encapsulates one

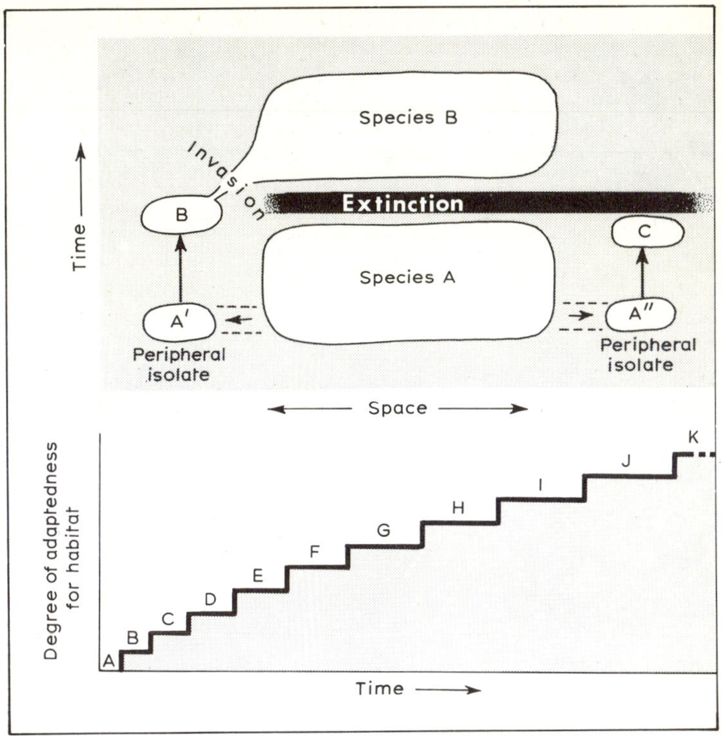

A graphical representation (top) of speciation and punctuated equilibria. Species A is replaced by species B, which evolved rapidly from the isolate A'. The lower graph shows the decrease in rate of species turnover and corresponding slowdown in rate of evolution. Species A is the first invader after a mass extinction. It is poorly adapted and is replaced by B. Each species is better adapted than the last, and so lasts longer

of the most prominent observations from the fossil record—species turnover. Each species of mammal-like reptile that has been found appears suddenly in the fossil record and is not preceded by the species that is directly ancestral to it. It disappears some time later, equally abruptly, without leaving a directly descended species, although we usually find that it has been replaced by some new, related species. The concept of punctuated equilibria—which envisages evolutionary change occurring in a series of jumps, with relatively little change in between—was introduced in 1972 by Niles Eldredge and Stephen Jay Gould, and accounts for the observed pattern rather well.

According to this concept, intermediate stages between known species are not found in the record because most evolutionary change occurs in very small, geographically isolated parts of the main species population. Such a peripheral isolate, as it is termed, can evolve very rapidly, for three main reasons. It includes only a few individuals, it is isolated from the main gene pool of the species, and it inhabits an environment different from that of the rest of the species. The large original population of the species is at the same time highly susceptible to extinction; because its population is large its potential rate of evolution is low and therefore it cannot change quickly in response to a minor change in the environment. In the course of time, such a species would become extinct and its place in the habitat would be taken by a related species, newly developed from one of the peripheral isolates. The fossil record shows this as the sudden replacement of one species by another within a lineage, rather than the gradual change of one species into another.

Extinction, then, is the normal, likely response of species to a relatively small environmental change, and one can speak about the rate of extinction of the species that make up a lineage, as well as about the rate of production of new species. If the rate of speciation is persistently greater than the rate of extinction, the number of species in the lineage will increase—an adaptive radiation. Conversely, if species become extinct more often than they arise the number in the lineage would decline, eventually to zero—a mass extinction (Rule 1).

Rates of speciation and extinction must depend on random events to some extent, but the fossil record also suggests that the rate tends to be different in different ecotypes (that is, in animals adapted to ecologically different habitats). For example, because speciation depends on parts of the main population becoming isolated, we would expect that small animals—which can be isolated by less dramatic geographical features—would be more likely to speciate than large animals. Perhaps, too, insectivores, with a diet that is typically rather cosmopolitan, might be more resistant to extinction than the average herbivore or large carnivore, with more specialised diets. So—as in Rule 8—the forms most likely to survive a mass extinction are small carnivores (which includes insectivores) followed by small herbivores. And similarly in the minor mass extinctions (Rule 7), it is larger types that are most at risk.

The speed at which morphology—form—changes will depend both on the rate of speciation and on the size of change that accompanies each new species. There has been a lot of speculation recently about whether larger changes than normal can occur during speciation. The fossil record of the mammal-like reptiles can do no more than suggest that such macromutational change is possible. What it does show clearly, however, is that the overall rate of change can vary enormously under varying circumstances. Immediately after a mass extinction the main lineages of the new radiation appear in the fossil record more or less fully developed (Rule 3). One possible explanation is that after the mass extinction had removed most of the species in a habitat, relatively ill-adapted species from the periphery were free to invade the range and establish themselves as new species populations. These new species in turn gave rise to their own peripheral isolates, and hence to a further generation of species. Because the present occupiers of the range were relatively ill-adapted some of the new peripheral isolates would be highly likely to form species better suited to the habitat, so that they would replace the existing species. This cycle of events would continue, with each new generation of species being better adapted than the one it replaces. This being so, it becomes less and less likely that a new generation of species will arise that is even better adapted.

The outcome of this is that species turnover is rapid at first, but falls as the species become better adapted. If the earlier phase of this programme (say the first dozen or so species) took a brief period of time (say a few thousand years), then the chance of us finding these forms among the fossils is very low compared with the chance of finding representatives of the later part of the programme, when the members of the lineages were fully adapted, definitive types that lasted for a far longer time.

Although we can see certain trends, we will probably never know what exactly caused morphological change in the synapsids to go in the particular directions it did, but there are certain interesting lines of investigation. Most obviously, there are the ecological opportunities available. Thus a normal terrestrial habitat offers opportunities for carnivores and herbivores of various sizes and perhaps a number of more specialised types such as scavengers. So it is only to be expected that each phase of the radiation followed a broadly comparable course, with similar ecotypes in each case (Rule 5).

More interesting is Rule 2—that the radiation begins from a small carnivore. Small herbivores also survive mass extinctions, so why are the herbivores of the new generation not derived from these? Why, instead, are they remodelled from the carnivore lineage? One possible explanation is that after the mass extinction there is a period when there are no large carnivores around. The surviving herbivores are temporarily free of predation and without this selective pressure they lose their ability to avoid being captured. When, after a period, larger carnivores arise from the small survivors, they are able to wipe out the susceptible herbivores, who rapidly become extinct. This leaves the reptilian plant-eating niche open, to be filled by derivatives of the small carnivores.

The third rule that addresses the direction of morphological change is Rule 4, the progress towards mammalian structure. We can chart the acquisition of mammalian characters by studying those particular fossils that lie closest to the hypothetical lineage from primitive pelycosaurs to mammals. Each biological system that shows in the skeleton—primarily feeding and locomotion but also such things as hearing, respiration and brain size—evolved step by step towards the mammal condition. Since all the systems did this, there is an apparent correlation among them, although this correlation is not absolute because we do sometimes see that one system becomes more mammal-like without the other systems accompanying it. Such a pattern of evolution of the characters is called a correlated progression, and it comes about because all the structures and functions of an organism are integrated within the whole. They must, perforce, change together if the animal is to continue to work properly.

The mammal-like reptiles were terrestrial animals and so faced the three major problems of life on land: temperatures that fluctuate wildly, water loss and gravity. But an animal cannot solve any one of these problems without aggravating the others. For example, if an animal reduced water loss with an impermeable skin, it would no longer be able to use evaporative cooling—sweating—to prevent overheating. Conversely, the animal can isolate itself from the harmful effects of temperature fluctuation by evolving a higher and constant body temperature, but only at the price of aggravating the problem of water loss.

It is clear that these problems (along with various lesser ones that face a land animal) could only be solved together, in an integrated fashion. A modification that bears on one of the problems would have to be accompanied by appropriate modifications bearing on the others. Translated into evolutionary terms, any one functional system could change only a small amount and would then be forced to wait, so to speak, for the other systems to catch up before it could take another incremental step.

Modern mammals have largely overcome the problem by refining the mechanisms that maintain their bodies in a constant state, but all the parts and processes involved are intimately linked.

The progress towards mammalness that we see in the evolution of the mammal-like reptiles must therefore be interpreted as a gradual improvement in the level of homeostasis—that is, in the extent to which the body is maintained in a state of constancy—associated with a gradually developing independence of fluctuations in the environment. Homeostasis is probably a virtue of any land organism, and its evolution is therefore manifested in all the ecotypes of mammal-like reptiles.

The fossil record of the mammal-like reptiles, to a greater or lesser extent in common with other fossil records, does contribute to evolutionary theory. It indicates two important phenomena that must be accounted for by any comprehensive theory of evolution, and that cannot be predicted from a study of living organisms alone. The first is species turnover, and the dynamic dimension it gives to an evolving lineage. The second is the part played by mass extinction, both as a filter that allows the differential survival of some lineages and not others, and also because of the events which follow a mass extinction. Conditions at such times seem to have been so different from normal that certain evolutionary events could occur which would not be predictable from a knowledge of the ecology of modern animals alone. Those times seem to have been the periods of maximum evolutionary change and were therefore of the greatest consequence in determining the whole course of evolution.

PART IV
PATTERN IN EVOLUTION

The pattern of relationships between organisms are what one would expect of Darwin's process of descent with modification. But are there any independent tests of the theory? There are, as these three chapters reveal

Chapter Ten

Cladistics and classification

Taxonomy has always been a dual-purpose weapon: it organises knowledge and reflects relationships. These two purposes sometimes come into conflict

COLIN PATTERSON

CHARLES DARWIN found one of the strongest arguments for his theory in the fact that animal and plant species fall into groups, and that these groups form a nested series, or hierarchy. Smaller groups of species, such as owls and ducks, or seals and deer, are included within successively larger groups: birds and mammals within vertebrates, vertebrates and invertebrates within animals, and so on. But hierarchical classifications were not Darwin's invention; the one we use today is derived from the work of the 18th-century Swedish naturalist Carl von Linné, or Linnaeus.

Linnaeus introduced his system partly as a convenient aid to memory, a means of making comprehensible the diversity of nature. But Linnaeus also had a higher purpose than merely to catalogue nature. He believed that he was uncovering the plan of the Creator. Linnaeus and his successors recognised genera, families and other categories on the basis of similarities in structure, and believed that each group had a set of features which were its essence, or ideal plan, corresponding to something in the mind of the Creator. The science of comparative anatomy developed as a means of searching out these ideal plans, or archetypes.

By the time Darwin published *The Origin,* Linnaean hierarchical classification and classical comparative anatomy were highly developed. Darwin's contribution was to suggest, by detailed argument, that the relationship between the species of a genus, or genera of a family, is a "blood" relationship, caused by descent and divergence from a common ancestor. The Darwinian interpretation of a Linnean hierarchy is that a classification based on similarities and differences between species might accurately reflect the relationships of common ancestry between them (Figure 1). He wrote in *The Origin:* "Our classifications will come to be, as far as they can be so made, genealogies; and will then truly give what may be called the plan of creation".

Darwin's expectation has not yet been fully realised. This is partly because classifications today, as in Linnaeus's time, have two purposes—to express evolutionary relationships and to act as *aides-mémoire* or simple summaries of knowledge. These two aims come into conflict because relationships of common ancestry are almost invariably more complicated

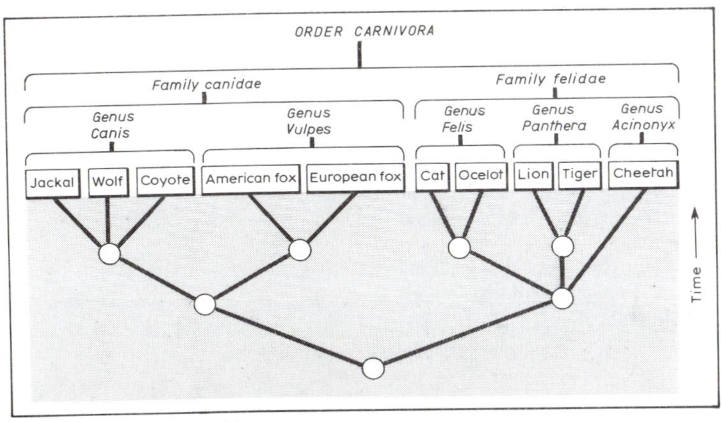

Figure 1 *Classification and evolution. The boxed names are living species, and the names above show how they are classified within the Linnean hierarchy. Species nest within genera, genera within families, and families within orders. The lower half depicts a Darwinian interpretation; the relationship is phylogenetic, the result of descent from common ancestors (open circles) more or less distant in time*

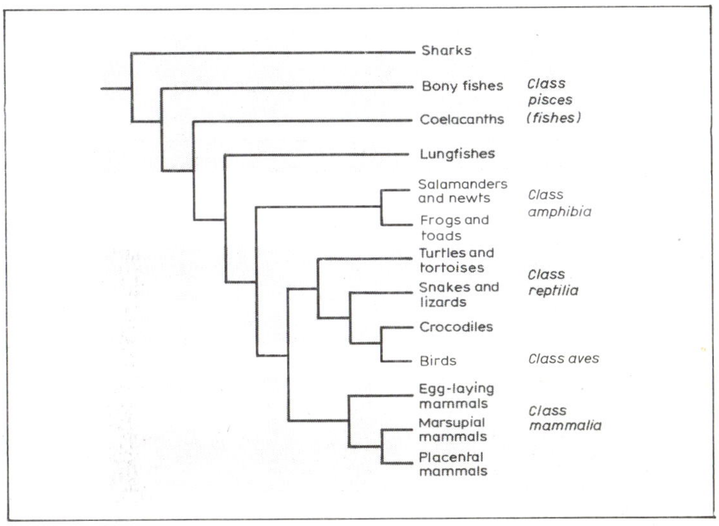

Figure 2 *Tradition and ancestry. On the right are the traditional classifications of vertebrates, and on the left the ancestry of the same animals. In the genealogy, the groups are indicated by vertical bars; only two of these—amphibians and mammals—correspond to the taxonomic groups. All the other taxonomic groups contain either too few—birds—or too many—fishes—members*

than the relationships of similarity and difference on which Linnean classification is based (Figure 2). Cladistics (from the Greek for "branch") is one method of biological classification that offers a new approach to the problem. As a method, cladistics has become both popular and controversial during the past 10 years or so; a review of the history of cladistics may help to explain the popularity and the controversy.

The basic principles of cladistics were set out by the German entomologist Willi Hennig in a 1950 book, but his ideas received little attention outside continental Europe until 1966, when an English translation of a much modified version was published as *Phylogenetic Systematics.* The keys to Hennig's system are his definitions of phylogenetic relationship and his discussion of how relationship is recognised. Hennig defined relationships by branching diagrams; hence "cladistics". He saw these as evolutionary trees, and argued that relationship in his sense, closer common ancestry, was evinced only by shared derived characters (which he christened *synapomorphies*) inherited from that ancestor. Shared primitive characters (*symplesiomorphies*), inherited from a more remote common ancestor, are irrelevant or misleading in the search for phylogenetic relationship: for example, in Figure 3 the relationship between species *C* and *D* may be obscured or masked by features inherited by only one of them from the ancestor common to *A* and *B* (at t_1). In the same way, the teeth and long bony tail of *Archaeopteryx* are irrelevant to its relationship with birds, because lizards, salamanders and fishes also have teeth and long bony tails. Feathers and a wishbone (furcula), however, are shared derived characters (synapomorphies) unique to *Archaeopteryx* and birds. A third type of character, those unique to any one group, Hennig called *autapomorphies.*

The three terms autapomorphy, synapomorphy and symplesiomorphy describe the distribution of characters relative to a particular problem: feathers are an autapomorphy of birds when one is interested in what other (featherless) groups are most closely related to birds, a synapomorphy when one is interested in the relationships of *Archaeopteryx,* and a symplesiomorphy for relationships within living birds.

From these ideas, Hennig derived definitions of three types of group. Monophyletic groups are those containing *all and only* the descendants of a common ancestor (*ABCD, BCD* or *CD* in Figure 3). Monophyletic groups are characterised by shared derived characters or evolutionary novelties: Mammalia and Vertebrata are examples with their fur and backbones respectively. Para- and polyphyletic groups are those which do not contain all the descendants of a common ancestor. A paraphyletic group is left behind when one or more parts of a monophyletic group are removed because the members share *derived* characters. Paraphyletic groups are characterised by shared *primitive* characters: Reptilia and Invertebrata are examples but by their very nature share no distinguishing character. There is nothing that defines a reptile in the same way that defines a mammal. Polyphyletic groups are those whose defining features are inferred not to have existed in their common ancestor. Polyphyletic groups are characterised by convergent characters: a group of all winged animals (birds, bats, insects) would be an example.

One source of controversy is evident here, for cladists deny that paraphyletic groups have any place in systematics, whereas such groups are traditional and popular, especially amongst palaeontologists, whose "ancestral groups" are all of this type: examples of paraphyletic groups among vertebrates include Pisces (fishes), Amphibia, Reptilia, and many smaller or extinct taxa. Harvard biologist Ernst Mayr accused Hennig and his supporters of misusing well-known words such as "relationship" and "monophyly". He proposed that relationship be defined as "genes in common" or "inferred amount of shared genotype" (see Figure 5), and that monophyly be defined not in Hennig's sense, as *all* descendants of a common ancestor, but in a less restrictive sense, by immediate common ancestry alone. By this definition, all possible groupings in Figure 3 could be monophyletic. The intention of Mayr's definitions was to permit paraphyletic groups, in order to avoid a second controversial aspect of Hennig's ideas, his method of forming a classification.

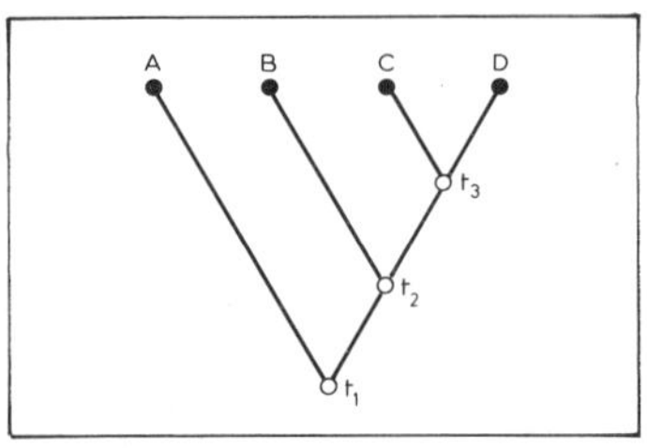

Figure 3 *Hennig's definitions. Two species (*C *and* D*, for example) are more closely related to each other than to a third (*A *or* B*) if they share an ancestral species (at t_3) not shared with the third*

For Hennig, a classification should name and rank all monophyletic groups and only those groups, and should give equal rank to "sister groups", the name he gave to species or groups which are each other's closest relatives: in Figure 3, *C* and *D* are sister groups, and *B* is the sister group of *C+D.* The motto of Hennigian systematists is "search for the sister group", because any species or group should have one species or group to which it is most closely related. Cladistic classifications are thus direct translations of phylogenetic diagrams.

Answering such critics as Mayr was one of the stimuli behind more recent developments in cladistics, leading to what has been called "transformed cladistics". As his title *Phylogenetic Systematics* implies, Hennig's work was set in an evolutionary framework. His definitions of relationship, of monophyly and paraphyly, and of symplesiomorphy and synapomorphy, were based on common ancestry and the notion of evolutionary change. Hennig's branching diagrams were evolutionary trees, with an implicit time axis, and with forks denoting splitting of ancestral species. But it is possible to look at diagrams like Figure 3, and at the definitions Hennig associated with them, in a more general framework, one which has no evolutionary implications. Branching diagrams can be seen not as evolutionary trees, but as cladograms in which there is no time scale and the nodes imply shared characters (synapomorphies) rather than common ancestry. This distinction, between cladograms and trees, may seem like hairsplitting, but it is important. A cladogram is a summary of pattern, the pattern of character distribution, or of hierarchy in nature—what pre-Darwinians called "the natural hierarchy". An evolutionary tree is a summary of pattern with an added summary of process, the historical process of descent with modification that caused the pattern of characters. In a tree, the nodes are real ancestors (though they may not be identifiable), the forks are speciations, and the lines of the diagram are lineages of descent with modification. In a cladogram, the nodes are shared characters and the lines are immaterial—one can as well represent the pattern by a Venn diagram (Figure 4).

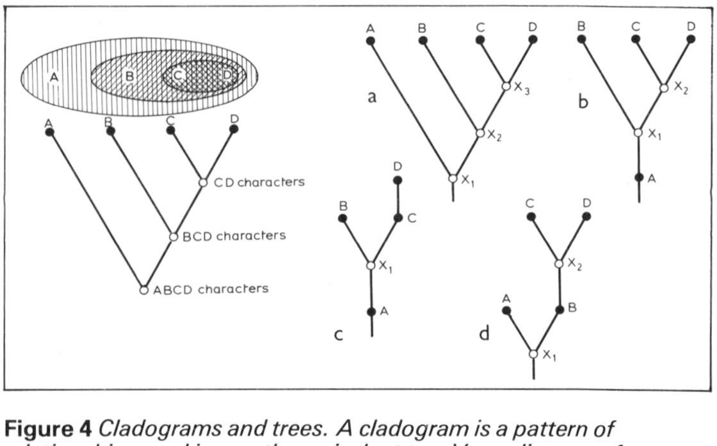

Figure 4 *Cladograms and trees. A cladogram is a pattern of relationships, and is exactly equivalent to a Venn diagram of sets and subsets (left). Trees (right) carry an added implication of evolution and time. The 4 trees are simply representatives of the 12 possible trees one could construct from the cladogram*

Thus a tree says much more than a cladogram. Tree **a** in Figure 4, for example, says that none of the four species is ancestral to any other, that the three ancestral species are unknown, and that speciation was dichotomous, with the unknown ancestral species becoming extinct on each occasion. The other trees offer other evolutionary scenarios. In short, cladograms are less restrictive, or more general, than trees, because a cladogram is equivalent to, or summarises, a set of evolutionary trees. And a cladogram, as a summary of pattern, is not necessarily evolutionary.

But there are still evolutionary implications in the notion of advanced characters (synapomorphies) and primitive characters (symplesiomorphies). If these categories of character must be distinguished before the cladogram is constructed, then it seems necessary to understand evolution before making a classification, and there is no sense in which the cladogram is independent of evolution. But this is not so. Methods of distinguishing primitive and derived characters have been much debated, and the lack of foolproof procedures has been one of the arguments used by critics of cladistics. However, there is one apparently foolproof method: it depends not on "evolution" as we understand it, but on the usage of that word among pre-Darwinian biologists. To them, "evolution" meant the unfolding of pattern, increase in complexity, or differentiation seen in the development of individual organisms—the process of embryology rather than of descent with modification.

Embryology is the only direct evidence we have of transformation of form. We can observe that all organisms begin life (in the egg, spore or seed) without certain features, that some organisms (a group) then develop those features in a general or "primitive" form, and in some smaller set of organisms (a subgroup) those features are refined into a particular or "specialised" form. This applies to a great variety of features—to structures such as the tail, limbs and brain of vertebrates, the leaves of plants, and so on, and also to features of physiology or behaviour. Such observations were the basis of a law proposed in 1828 by the pioneering embryologist K. E. von Baer: "In development the general characters appear before the special ones". That law provides the framework of the systematic hierarchy, groups and subgroups circumscribed by recognition of general (primitive in evolutionary terms) and special (derived or advanced in evolutionary terms) characters. In this light, the synapomorphies and symplesiomorphies of phylogenetic systematics are exactly equivalent to the homologies of classical biology. Every homology characterises a group at some level in the hierarchy, and symplesiomorphy and synapomorphy are terms for homologies that stand in hierarchic relation: a symplesiomorphy (general character) makes a group, and a synapomorphy (special character) makes a subgroup. In this light, too, the argument between cladists and evolutionary systematists comes into clearer focus.

That argument centres on the kinds of group that should be recognised in classifications. Cladists demand that groups be characterised by synapomorphies (or homologies), so that they are monophyletic (or natural). Evolutionary systematists agree that cladistic analysis is the best way of approaching systematics, but in classifications they wish to take account of rate or degree of evolutionary divergence—*anagenesis*—in addition to the branching or *cladogenesis* recognised by cladists. Evolutionists regard production of classifications as a two-step process, cladistic analysis followed by analysis of divergence. For instance, when one group differs excessively from its sister group and other close relatives (*D* in Figure 5), evolutionists wish to recognise the inferred rapid divergence by raising the rank of that group. Well-known examples include man (family Hominidae) versus our sister group (chimpanzees and/or gorillas—genus *Pan* or subfamily Paninae), birds (class Aves) versus their sister group (crocodiles, order Crocodilia), and land vertebrates (Tetrapoda, ranked above class) versus their sister group, lungfishes (Dipnoi, ranked below class). An inevitable consequence of the structure of the hierarchy is that when this is done the sister group of lower rank (chimps and gorillas, or crocodiles, or lungfishes) is included with others in a paraphyletic group equal in rank to the divergent group. Thus chimps and gorillas are placed in the family Pongidae, which also includes *Pongo*, the orang-utan, and sometimes also the gibbons; crocodiles are placed in the class Reptilia, which also includes lizards, snakes and turtles; and lungfishes are put in the class Osteichthyes, with other bony fishes.

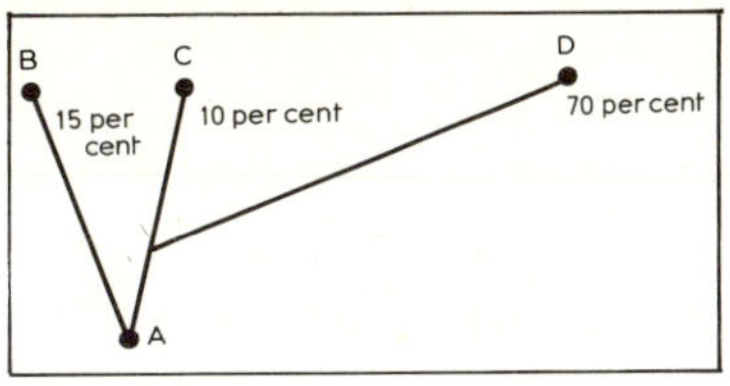

Figure 5 *Mayr's definitions. In this tree the slope of the branch corresponds to the rate of evolution, and the percentages represent the degree of difference from the ancestral species* A. *Mayr would classify* C *with* B *because the two are more like their common ancestor. A cladist classifies* C *and* D *together because they share a common ancestor that* B *does not*

A further consequence is that the cladogram reconstructed from the evolutionary classification disagrees with the cladogram based on character analysis (Figure 6). Evolutionists justify their strategy by asserting that classifications must be grounded in evolutionary theory, and by appealing to the purpose of biological classification. One may argue about the function of classification, but it is generally agreed that the greatest virtue of a classification is predictivity, or congruence with the greatest number of characters. John Stuart Mill's doctrine is often quoted: "The ends of scientific classification are best answered, when the objects are formed into groups respecting which a greater number of general propositions can be made, and those propositions more important, than could be made respecting any other groups into which the same things could be distributed". The "objects" or "things" in biology are species, and the sense of Mill's statement is that new information, or different types of characters, will conform better to the adopted classification than to another.

Thus evolutionists argue that new information—newly discovered or investigated characters—will conform to the groupings produced by taking anagenesis, or degree of divergence, into account. In the hominoid example, they predict that characters of chimpanzees and gorillas will be shared with orang-utans rather than with humans. During the past decade or so, newly discovered characters bearing on this example have come mostly from the molecular field, particularly amino-acid sequences of proteins and, more recently, nucleotide sequences of RNA and DNA. Without exception, these genetic details fail to meet the prediction. The best test yet is a sequence of 896 nucleotides from the mitochondrial DNA of man, chimpanzee, gorilla, orang-utan and gibbon: 284 of these nucleotides differed among the five species, and whereas conventional classification predicts that chimp and gorilla should be more like orang, nucleotides unique to man, chimp and gorilla outnumber those unique to chimp, gorilla and orang more than fourfold.

Since the "predictive" defence of evolutionary classification fails, some evolutionists have offered another, that conventional classification is preferable because it is more stable, and not subject to frequent changes as ideas on the interrelationships of species fluctuate. To stick to the hominoid example, this defence implies controversial or newly-proposed notions of the interrelationships of apes and man. But, as Figure 7 shows, the cladogram derived from molecular data on apes and man agrees with the first detailed evolutionary tree ever published, Ernst Haeckel's in 1866. It also agrees with T. H. Huxley's words in *Man's Place in Nature* (1863): "It is quite certain that the Ape which most nearly approaches man, in the totality of its organisation, is either the Chimpanzee or the Gorilla". In the same work,

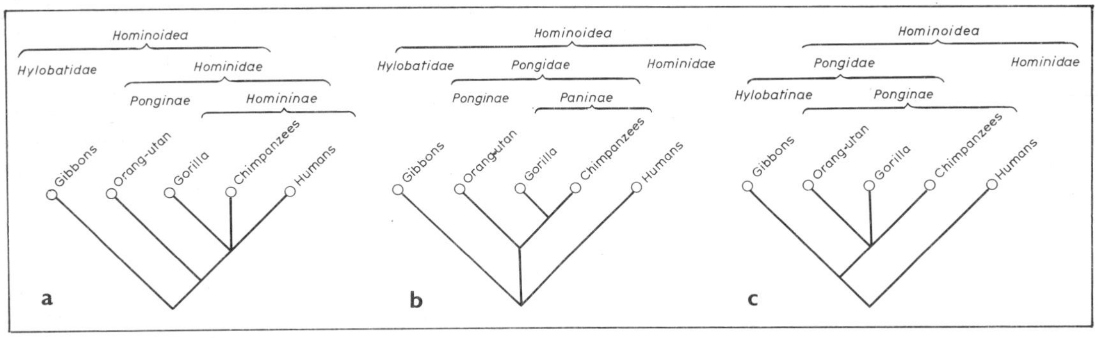

Figure 6 *Cladograms and classifications of hominoids. The ancestry of man and apes, as currently understood, is shown in* a*, with the classification that this cladistic analysis would support. Two classifications in use are shown in* b *and* c*, with the cladistic relationships they imply*

Huxley epitomises the predictive consequence: "Whatever part of the animal fabric . . . be selected for comparison—the result would be the same—the lower Apes and the Gorilla would differ more than the Gorilla and the Man". Is stability worth more than a century of conflict with evidence?

Paraphyletic groups may seem a small price to pay for the advantages claimed for them—convenience, stability, and the recognition of evolutionary advances. Yet groups such as the Pongidae and Reptilia, necessitated or permitted by evolutionary classification, have one crippling disadvantage, and unfortunate consequences. The disadvantage is that those groups are uncharacterisable: they have no characters of their own. The only statements that can be made about them are negative, that they lack the characters of the groups that have been detached (birds and mammals detached from reptiles; humans from apes). If, to quote John Stuart Mill again, we view "classification as a logical process subservient to the investigation of truth", or if, in more prosaic terms, "systematics in general consists of the search for defining characters of groups" (as Gareth Nelson and Norman Platnick put it), paraphyletic groups deny both propositions; the "truth" of human/ape relationships has been around for well over a century, and the defining characters of groups like Pongidae and Reptilia are non-existent. Those groups have no real existence: they are artefacts.

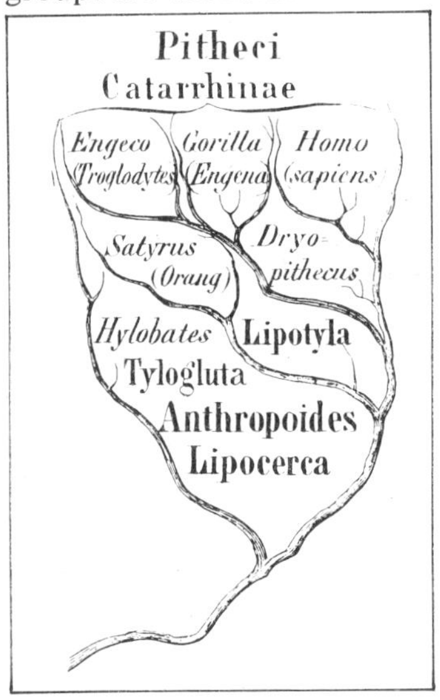

Figure 7 *Haeckel's evolutionary tree, published in 1866.* Engeco *is chimpanzee,* Hylobates *gibbons. Note the similarity to Figure 6*a

And those artefacts have unfortunate consequences. Historical statements that are superficially objective, like "man is descended from apes" or "birds are descended from dinosaurs", *always* invoke such groups. The concept of ancestry, as part of evolutionary theory, has so captivated systematists that they are trained to think in terms of trees, of ancestry and descent. But the only clues to relationship between species are uniquely shared characters (homologies). Such homologies may be seen as evolutionary novelties originating in the ancestor unique to those species, yet shared characters alone will not distinguish between the two types of evolutionary relationship, direct ancestry and common ancestry. Two species may share a character either because one is the ancestor of the other, or because both have a common ancestor (Figure 8).

So, therefore, hypotheses of ancestry must depend on some justification beyond character distribution: for instance, that the ancestor be represented by fossils occurring earlier in time than the descendant, and that the fossil record should be sufficiently complete and well-sampled for there to be no other potential ancestors. But fossils, too, present peculiar problems in analysis. They are very incomplete when compared with living organisms, so it is generally impossible to be sure just what characters they had; they are decipherable as organisms only by comparison with living relatives, which means that they must be assigned to extant groups before they can be properly interpreted; species limits and geographic ranges of fossils are largely guesswork; and so on.

Because of all these problems, it is rare to find palaeontologists offering ancestral species, or doing so with any conviction. Instead, they usually propose "ancestral groups", as approximations to the truth, with the claim that the true ancestor, if found, would fall within the group. Extinct ancestral groups are paraphyletic, just like Reptilia and Pongidae, and have the same status as uncharacterisable artefacts, with the added complication that the included fossils may not be complete enough to say even what characters they lack. Yet these flawed artefacts play a central role in phylogenies—accounts of the evolutionary descent of lineages. This raises yet another problem, for groups cannot evolve—species are the largest units capable of change. Thus cladistics calls into question much of conventional evolutionary history.

Ancestors, either species or groups, have a place in trees but not in cladograms (Figure 4). Cladists recommend avoiding some of the problems with fossils by treating them in the same way as living species, as potential twigs of cladograms rather than stems of trees.

A further disadvantage of paraphyletic groups is that they are readily mistaken for real (monophyletic) groups. Staying with the example of man and apes, within the past few years I have seen publications by specialists dating the separation of man from apes at times that range from less than 5 to more than 75 million years ago. Any one of these dates *might* be true—we have, as yet, no foolproof way of knowing. But it is noticeable that the more remote dates are accompanied by diagrams in which the gibbons, orang-utan and African apes are shown as monophyletic (as in Figure 6**c**), while the middle range of dates is found with diagrams in which the orang-utan is related to African apes (as in Figure 6**b**): in both instances, the specialists have treated the Pongidae as a monophyletic group. Had they recalled T. H. Huxley's words and the cladogram they imply (Figure 6**a**), they might have seen the implication that whatever date is assigned to the man/ape split, the chimpanzee/gorilla split will be of comparable age, and the orang-utan should diverge even earlier. It was mistaking paraphyletic groups for monophyletic groups that prompted these experts, for whatever reasons, to drive the human lineage so far into the past.

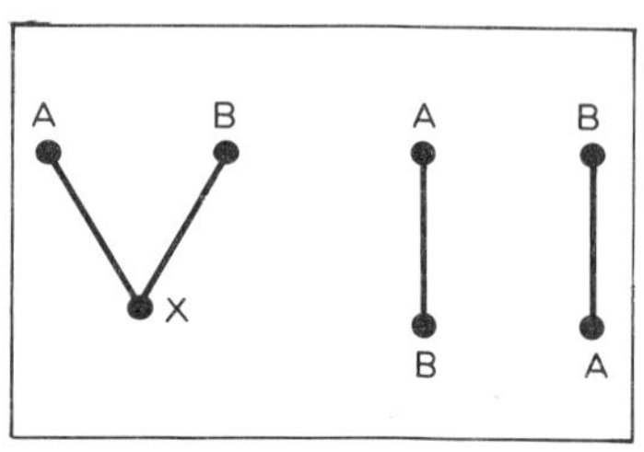

Figure 8 *Shared characters alone tell us little about descent; and B could both be descended from some common ancestor X, or one of them may be descended from the other*

The three components of the history of life are form, time and space. The biological disciplines dealing with these are systematics, which concerns the variety of form, palaeontology, concerned with that variety in time, and biogeography, concerned with that variety in space. Cladistics calls into question traditional attitudes in all three, and offers a new approach to comparative biology which has a coherent theoretical base that is not necessarily tied to evolutionary theory. As a science of pattern, cladistics holds out the possibility of a reconstruction of the history of life in space and time that does not depend on Darwinian or neodarwinian presuppositions. The interest of that reconstruction or cladogram is that theories of process—neodarwinism or any other—can be tested only against nature, and the best test will be their success in explaining past and present configurations of life. But if we are taught, as we have been, to see that pattern through the spectacles of evolutionary theory, how could the pattern ever test the theory?

Chapter Eleven

Proving the pattern of life

The structure of animals, their anatomy, tells us much about their relationship one to another. The structure of their body proteins can tell us even more

JEREMY CHERFAS

EVOLUTIONARY biology has always had two principal goals; to understand the processes that lie behind evolutionary change and to trace the pattern of life on Earth. Darwin demonstrated that hierarchical taxonomic systems, especially the binomial system of Linnaeus, were just as one would expect of a process of evolution by modification and descent (see chapter ten). Because evolution was a continuously diverging process, living things would be related to the extent that they shared a common ancestor. Those most recently diverged from a common ancestor would be more similar than those that had parted company in the more distant past. There is only one true set of relationships among living things, and the problem became to reconstruct that tree from the organisms themselves.

Early taxonomists relied a great deal on comparative morphology, seeking similarities and differences that would reveal which organisms were more closely related and which more distant. The data were measurements, impressions, feelings about the species to hand, and the methods used to derive an evolutionary tree from these data largely intuitive. Gradually explicit rules for building evolutionary trees from the raw facts of similarities and differences were developed, but two outstanding problems remained unsolved. One was the question of subjectivity in the selection of facts. It was quite possible for a taxonomist to attach more weight to those features that supported his prior position and to discount those that did not. The second problem relates to the first; given two alternative trees, how could one decide between them? Again, opinion counted for a great deal. It was always possible to find some reason to prefer one tree over another, but there existed no way of demonstrating unequivocally that the preferred tree was indeed closer than the rejected one to the true tree. Molecular biology changed all that.

The outward form of an animal, so important in deciding relationships, is the end product of a complex web of interactions. At the centre of that web is the nucleic acid DNA, which stores the information needed to construct any given animal. It is differences in the DNA that ultimately underlie heritable differences between animals, and changes to the DNA—mutations—cause changes to the animal. Members of the same species can interbreed and share their DNA with one another. Any changes that arise in the DNA can then spread through the population. Animals in different species cannot breed and so cannot share DNA. Any changes that arise in one of them will be confined to that population and its descendants. So as soon as two species begin to separate from a common ancestor, their genetic messages begin to move apart. DNA thus becomes another character for the taxonomist to use. More similar DNA implies more recent common ancestry. And DNA, while it may be at the centre of the web, is not the only molecule that can be studied. The proteins that DNA codes for and the antibodies that the immune system generates in response to foreign proteins may be removed from the DNA itself, but they too can be used as taxonomic characters.

Molecular phylogeny—the use of biochemicals to illuminate relationships—is not new. It began at the turn of the century with George Nuttall, professor of biology at Cambridge University. Nuttall used the newly-discovered immune reaction to investigate the degree of "blood relationship" between various species, and came up with results that were in broad accord with the trees devised by more conventional methods. Nuttall was followed by others, notably Morris Goodman at Wayne State University in Detroit, who refined his immunological methods in the early 1960s. Goodman demonstrated that man, chimpanzee and gorilla were one another's closest relatives. He found it easy to separate any one of them from the orang-utan, but could not say which pair of the threesome was closest. This was a profound disagreement with the phylogeny that had grown up around man, which placed all the great apes as a group distinct from man; on that score chimpanzee and gorilla should associate with one another and with the orang, but not with man. (Ironically, Goodman's "new" phylogeny represented a return to the tree that the early anatomists, including Thomas Huxley, Charles Darwin and Ernst Haeckel, had constructed.) Many people refused to accept Goodman's results *because* those results disagreed with the conventional tree, and there was no way to arbitrate.

Goodman focused his attention on our closest relatives. At about the same time Walter Fitch and Emanuel Margoliash, working at Northwestern University in Illinois, took a broader look at the larger tree of life. Biochemists interested in metabolism had determined the sequence of amino acids of the protein cytochrome *c* in 20 species. Fitch and Margoliash took the amino acid sequences and, knowing the genetic code, asked what was the minimum number of changes to the DNA that would be needed to account for the observed differences between the various species' cytochrome *c*. From these data they constructed an evolutionary tree that, once again, agreed fairly well with the traditional tree. There were, however, some disagreements. On the basis of its cytochrome *c* the turtle (a reptile) ended up much closer to the birds than to the rattlesnake, another reptile. More worrying, the primates as a group seem, according to cytochrome *c*, to have split off from the other mammals before the marsupials split from the placental mammals.

The encouraging thing about the tree based on cytochrome *c* was that it was in such broad agreement with the traditional tree. Leaving small discrepancies aside, molecular phylogenies

Francois Gohier/Ardea

The chachalaca—a confusing bird from Mexico

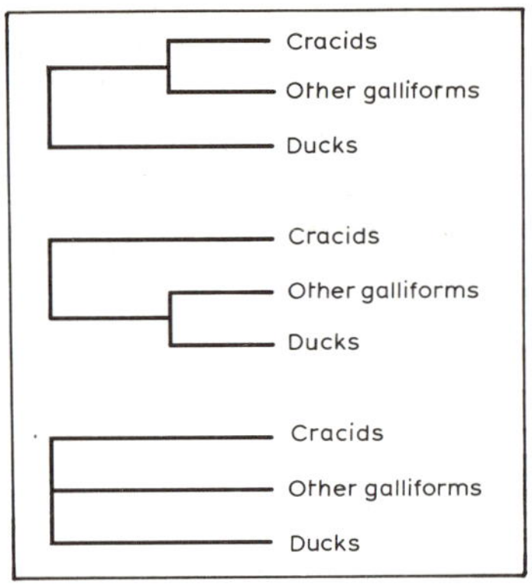

Figure 1 *Three alternative trees for the cracids. The morphologists (top) put them firmly with the galliforms. The molecules (middle) put ducks and galliforms together and separate from cracids. A compromise (bottom) has a three-way split, until better data come in*

reinforced the idea that there was indeed a single true tree. Molecular trees looked very like anatomical trees. What is more, other molecular studies provided collateral support. Different proteins—haemoglobin, lysozyme, albumin, ovalbumin, transferrin—resulted in very similar phylogenies. Indeed, biochemical investigations could now be used to settle long-standing questions of phylogeny that traditionalists could not resolve.

Vincent Sarich, working with Allan Wilson at Berkeley, provided one early example. Anatomists have long puzzled over the evolutionary status of the giant panda *Ailuropoda melanoleuca.* Some say it is nothing more than a bear. Others maintain that it is obviously related to the lesser panda *Ailurus fulgens,* and that as the lesser panda is closely related to the raccoons the giant panda is a raccoon. Yet a third position is that the giant and lesser pandas are related, but form a group of their own distinct both from bears and raccoons. In 1972 Chi-Chi, London Zoo's panda, died and Sarich got a sample of her blood. Two large proteins, albumin and transferrin, showed clearly that the giant panda was a bear, having only recently split from the line that leads to the other bears. Indeed the giant panda was more closely related to the black bear than to the lesser panda. The blood protein molecules agreed with the most thorough anatomical study, which concluded that "the giant panda is nothing more than a highly specialised bear", and the molecular method had begun to demonstrate its usefulness as a source of additional phylogenetic information.

Another exemplary piece of biochemical tree-building concerns the chachalaca *Ortalis vetula.* The chachalaca is a Mexican bird that has been classified, on morphological grounds, as a member of the order Galliformes, the order that contains the chickens and pheasants. It is placed in the family Cracidae, along with the guans and curassows, and that family is firmly a part of the galliform order. When Allan Wilson and Ellen Prager compared another protein, lysozyme *c*, of many birds they were surprised to get results that placed the chachalaca out on its own. The lineage leading to the chachalaca's lysozyme *c* was no closer to the chicken's than was the lineage leading to the duck's lysozyme *c.* And the duck is not even a galliform; it is a member of the Anseriformes. Again, the problem of deciding between alternative trees had arisen. The single molecule lysozyme *c* disagreed with the *Gestalt* of morphological features; which was correct (Figure 1)?

To answer, Prager and Wilson extended the molecular study. They looked at three more of the chachalaca's proteins; transferrin, ovalbumin and albumin. These proteins told exactly the same story as lysozyme *c.* The chachalaca lineage was only remotely related to the other galliform lineages. The relationship between the chachalaca and the chicken was about the same as that between the duck and the chicken.

Four separate proteins (and later a fifth), each of which can be considered to have evolved independently, give the same answer; the chachalaca is not really a galliform at all. Indeed, if anything it is further from the Galliformes than the duck, but because of the slight uncertainties in the method Prager and Wilson argued for a three-way split between ducks, cracids and non-cracid galliforms.

Two proteins for the panda, four for the chachalaca; slowly the molecular methods were becoming more powerful. The problem with proteins, especially when one measures not the protein itself but the immune response to it, is that so much potential information is wasted. Albumin is a big protein, about 600 amino acids long, and yet comparing two species' albumin you come up with a single arbitrary number, the immune distance between them. There is so much more information to be had. The amino acid sequence, as Fitch and Margoliash showed, provides a great deal more data, because each amino acid difference is potentially informative. The DNA sequence would be even better, because the redundancy of the genetic code means that the DNA can change with absolutely no effect on the protein it codes for. Knowledge of the DNA sequence in different species would easily provide enough information to perform statistical tests on alternative phylogenetic trees. This is exactly the approach that Allan Wilson has been pursuing; so as to clarify once and for all the relationships of most interest to us—our own.

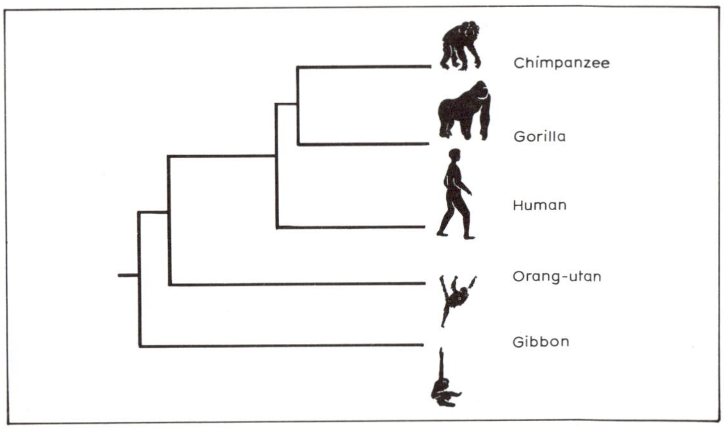

Figure 2 *The ape family tree, as established by a combination of molecular and anatomical evidence*

Ever since Goodman's discovery that man, chimpanzee and gorilla were impossible to distinguish immunologically, controversy has surrounded our relationship with the African apes. In 1967 Sarich and Wilson confirmed Goodman's finding and put numbers on the distances between the apes and man. That confirmation was based on the blood protein albumin. Over the next 15 years further confirmation dribbled in from other molecules, until in 1980 there were six independent proteins, all of which supported the grouping of man, chimpanzee and gorilla as a unit distinct from the other apes. All the molecular studies agree on this grouping. There are, however, still many obstacles. One is to persuade people that the molecules are telling the truth; despite the concordance between the various molecular studies and the best of comparative anatomy, there are those who resolutely maintain alternative phylogenies. In the face of this opposition, is there any chance that the molecular conclusions are mistaken? Another is to decide the exact branching order leading to man, chimpanzee and gorilla. Which pair shares a more recent common ancestor? Molecular phylogenies based on DNA, coupled with a rigorous statistical approach, have answered the first question unequivocally, and are close to a definitive answer to the second.

To get straight to the point, it is now possible to rule out a tree that associates the African apes with the orang with a certainty beyond the 0.0001 level of statistical significance. But the other important issue, the branching order for man, chimp and gorilla, is still a little doubtful. A brief association of chimp and gorilla after their line split from man's has the edge over the other possible trees by about 5 to 1. How have the molecules been able to provide such firm answers?

Spin-off from genetic engineering offered the means, and the interest was always there. Genetic engineers developed techniques to isolate specific segments of DNA and they also devised ways to examine the sequence of bases along the DNA. One, restriction mapping, takes advantage of so-called restriction enzymes to look for specific sequences on the DNA. These enzymes, isolated from bacteria, have the extremely useful property of being able to recognise a specific sequence of bases on the DNA. More than that, they cut the DNA at a particular point in the sequence. Any given restriction enzyme, then, will reliably chop a piece of DNA into the same collection of fragments every time, and those fragments can easily be separated and measured. There are now more than 300 restriction enzymes, many of which recognise different sites. By using several restriction enzymes, alone and in judicious combinations, one can build up a very detailed picture of a stretch of DNA, and that is exactly what Wilson, with Steve Ferris and Wes Brown, did.

The DNA they concentrated on was taken from the mitochondria of the apes. Each mitochondrion has its own piece of autonomous DNA roughly 16 000 bases long (which some people think is a remnant of the organelle's former existence as an independent bacterium; now it is responsible for respiration in the cell). Mitochondrial DNA evolves about 10 times faster than DNA in the nucleus; this means that there are 10 times more differences between the mitochondria of recently diverged species (which the great apes are) and offers the prospect of more information.

Wilson and his colleagues began with purified mitochondrial DNA from six species: gibbon, orang-utan, man, chimpanzee, pygmy chimpanzee and gorilla. They digested that DNA with 19 restriction enzymes, each of which recognised a different sequence of bases. Then they used the fragment sizes to construct very accurate linear maps of the six species' DNA. A close look at the six maps reveals the relationships among the species.

The first step is to get rid of any sites that are not phylogenetically informative, that is sites unique to just one of the six species. These would be consistent with any tree, so they are of no use. That leaves a total of 42 informative sites. Of those, 40 are identically placed in the chimpanzee and the pigmy chimp; they are the most closely related pair of the six, hardly surprising. The most distant pair are the gibbon and the orang-utan; they share just 19 sites.

The restriction maps can do far more though. They enabled Wilson to calculate the minimum number of mutational events that would be needed to explain the differences among the species, and then relate those differences to the various phylogenetic trees that could be drawn. The most parsimonious tree—the one that requires the fewest mutations to account for the distribution of restriction sites—places man, chimp and gorilla together and gives gorilla and chimp a common ancestor after they had split from man. This tree needs 67 mutations at the 42 restriction sites. To give the chimpanzees and man a common ancestor takes one more mutation, while a three-way split uses 71. (Putting all the great apes—orang, chimp and gorilla—together and separate from man also needs 71 mutations.)

Other restriction maps favour the same tree as the mitochondrial DNA. Liz Zimmer, a student of Wilson's, made an extensive map of the three genes that code for globins; 20 informative sites need 38 mutations to associate all the great apes but only 25 to exclude the orang-utan.

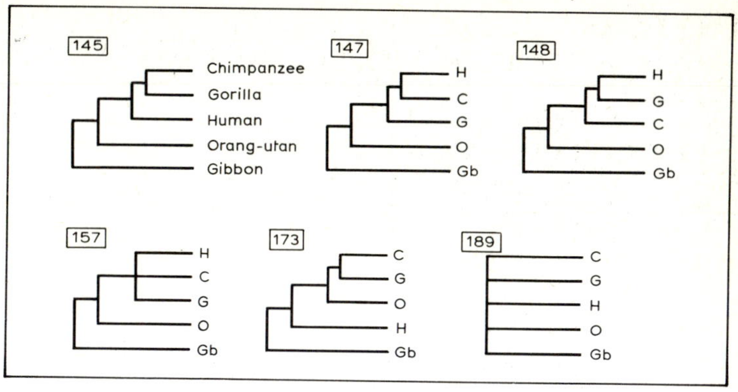

Figure 3 *Some alternative evolutionary trees for the apes. The number beside each tree indicates the number of mutations in the 896 base segment of mitochondrial DNA that would be needed to account for the particular branching order of that tree*

Even more information has come not from maps but from the actual DNA sequence. Wilson, with Wes Brown, Ellen Prager and Alice Wang, pulled a short sequence of 896 bases from the mitochondrial DNA of the five apes. The sequence codes for three transfer RNAs and for parts of a couple of unidentified proteins. Wilson and his colleagues discovered the identity of every one of the 896 bases for each species. The five sequences differ at 284 positions. All but one of these are simple substitutions of one base for another, but the orang-utan appears to be missing a base that the other four share. The closest relationship was between chimp and human, who had only 79 differences (9 per cent) and the most distant between gibbon and orang, who had 169 (19 per cent).

There are 90 phylogenetically informative sites in the DNA segment that Wilson examined. As with the mitochondrial maps, he could draw up various trees and see which one was most favoured. It was the same one that came out of the restriction maps, chimp and gorilla together briefly after a

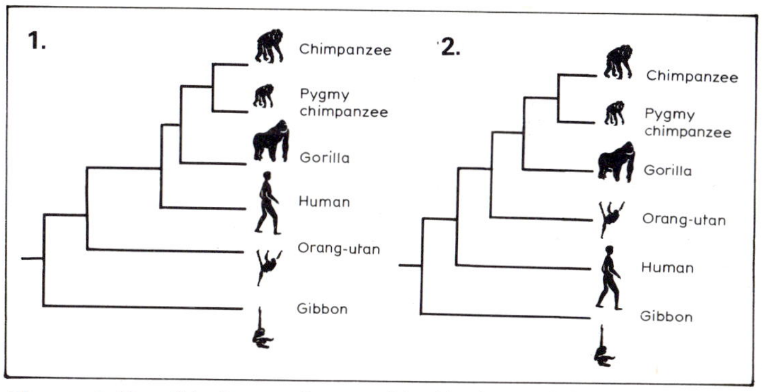

Figure 4 *Tree 1 is the one favoured by Huxley and the early anatomists. Tree 2 is an alternative favoured by most palaeontologists. The molecules, as the table shows, overwhelmingly favour tree 1 as being the correct tree.*

	Number of sites	Mutational events needed *Tree 1*	Mutational events needed *Tree 2*
Proteins	6	8	12
Restriction maps			
Mitochondrial DNA 1st half	21	32	34
2nd half	21	35	37
Globins	20	25	38
DNA sequences			
Mitochondrial *A*	23	39	42
B	22	39	44
C	23	31	40
D	22	37	48
TOTAL	157	246	295

split from man, and these three separate from orang-utan. This tree needed 145 mutations. The sequence data go beyond the restriction maps in one crucial respect. They make it very unlikely that the orang was ever associated with chimp and gorilla; such a tree needs 173 mutations, nearly a fifth more.

The sequence data provide less help in deciding the branching order of man, chimp and gorilla. An association of chimp and gorilla needs two fewer mutations than man and chimp, which in turn needs one less than man and gorilla. Interestingly, the three-way split that almost all previous molecular phylogenies have favoured seems to have been a reflection of the lower resolution of those other techniques. This pattern would need considerably more mutations, 157 as opposed to 147, in this stretch of DNA (Figure 3).

The particularly nice thing about the DNA data is that they are plentiful enough to allow statistical testing. Stimulated by the pioneering work of Alan Templeton of Washington University in St Louis, Wilson, Prager and Zimmer have divided the mitochondrial restriction maps into two subsets, and the sequence data into four. Add in the restriction maps of the globins and the amino acid information from the proteins that have been sequenced, and you have eight independent sources of information. Any tree can be checked against the data, and the number of mutations needed can be counted. The tree that needs the fewest mutations is probably the true tree (Figure 4).

It is a remarkable fact that all eight sets of data favour one tree over the other. "This," says Allan Wilson, "demonstrates the power and reliability of molecular trees". Furthermore, the molecular data distinguish clearly between a tree that has chimp, gorilla and orang on one branch and a tree that puts the orang on a separate branch; something, Wilson says, that "morphological analysis has not succeeded in doing".

There remains some doubt about the most recent branch, but the conclusion from the molecules, the most informative studies of our evolutionary history, is that the chimpanzee and gorilla shared a common ancestor for a brief period after they had diverged from man, and that the common ancestor of all three diverged from the orang-utan quite some time earlier. Any alternative is simply unsupportable.

Chapter Twelve

How to explain organic diversity

Natural selection can provide an explanation for the great diversity we see among living things, but if the explanation is to be correct we must take care in how we arrive at it

MARK RIDLEY

ONE HUNDRED years ago, Charles Darwin bequeathed to us a theory which is, in principle, able to explain every characteristic in every species. There are millions of different species, each of them with its own distinctive set of characteristics. To explain the patterns in the diversity of life—why birds and crickets sing, frogs croak, while other species are silent; why some animals are monogamous and take one mate for life, while others are polygamous; why some have two legs, others four—is a problem on a grand scale. Yet it is a problem that Darwin's theory would solve with only two ideas: history and design. Or, in everyday Darwinian language, phylogeny and adaptation. To account for some characteristic the Darwinian must first identify that part of the species' environment which makes the evolution of that characteristic advantageous. Then he knows why it has been favoured by natural selection.

Similarities and differences among species should, then, be explainable by similarities and differences of the species' lifestyle. This is the premise of the comparative method of explaining organic diversity. I can illustrate the comparative method by a textbook example. The body shape of the dolphin is similar to that of the shark, and many other kinds of fish. The comparative evidence in this example is that a certain body shape has evolved independently at least twice in fish and mammals that lead a particular lifestyle. Add some logical reasoning about how natural selection works, concerning the water resistance of different body shapes, the advantage of swimming faster and farther for catching food and avoiding being eaten. The conclusion is that the shape shared by shark and dolphin is an adaptation for swimming.

The argument divides into two parts. The reasoning about natural selection in this particular example is simple, but in many other cases it is easy to be mistaken about what natural selection can, and cannot, do. In recent years there has been a definite improvement in the standard of argument about natural selection, but this is not my theme. I am concerned with the second part, the comparative evidence.

Although the Darwinian explanation of organic diversity looks simple, it has proved too difficult, or perhaps just not interesting, for most of Darwin's followers. The last great Darwin anniversary was the 1959 centennial of *The Origin of Species.* I have looked through the many books commemorating the comfortable conferences of that year, but I met hardly any interest, and certainly no inclination to research, why organisms are as they are. The field has changed now, and since the early 1960s the analyses of adaptation have been setting the pace in evolutionary biology. But, despite our progress since 1959, we still do not have a full science of adaptation. The comparative method, which is one method of studying adaptation, still lacks a satisfactory technique with which to test among hypotheses. Although anniversaries are appropriate occasions to look back and complacently celebrate the past, I am not going to review the past 20, or 100, years. I am instead going to concentrate on a modern problem. I shall argue that a synthesis of ideas developed in that controversial school of taxonomy called cladism with all that is sensible in modern adaptationism can at last give an explanatory science of the diversity of life (see chapter ten).

In its naivest form the comparative method assumes that for each life style there is an appropriate set of adaptations. A practitioner would list many examples (the more the better) of similar adaptations, body shape in the example above, in species with similar life styles. Although many great biologists from Aristotle onwards have thus applied the comparative method, I think it is incomplete. It is really just an exercise in the selective use of examples. Our naive comparative biologist is like a lawyer, ransacking the annals of biology for observations that support his case but not obliged to notice the exceptions. A proper comparative method would have considered not only the shark and the dolphin, but also any species with a similar body shape but which are not active swimmers, and any active swimmers with different body shapes. These should be weighed against all species with different body shapes from the dolphin which do not move by swimming. Only when we know the number of animals in all four (or however many) categories can we see if there is a real

The cladistic method

Suppose that we are confronted by four species, and are studying a character called *A*. Two species, 1 and 2, have it in state *a*; the other two, 3 and 4, have it in state *a'*. The question is whether *a* or *a'* is the primitive character state in the group. One of the character states has evolved independently in the group, but was it *a* or *a'*? The cladistic method of outgroup comparison is derived from the assumption that convergent evolution is much less likely than similarity by descent from a common ancestor. The cladist examines a species related to the group and finds out whether it is *a* or *a'*. Whichever it is is likely to have been the primitive character state in the four species.

If *a* is primitive, *a'* must be scored as having evolved once; if *a'* is primitive then *a* has evolved once. In either case, further outgroups have to be examined before the independent evolution of the primitive character-state of these four species can be scored. Eventually we might come to an outgroup with state *a''* in which case *a* (or *a'*, as appropriate) must have evolved after divergence from the common ancestor with that outgroup.

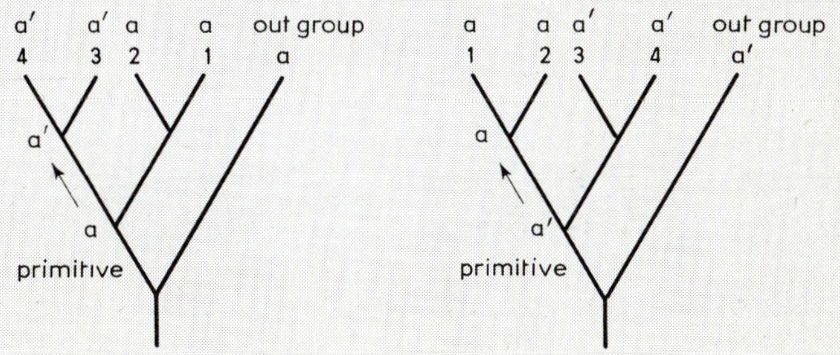

relationship between body shape and mode of movement. As a first requirement, then, the comparative method must be applied to *all* the facts; it is not good enough to select only those that confirm a theory. It might be thought that this is an obvious point, but I know of no attempt before 1960 that employs a systematic rather than a selective use of the comparative method.

Two of the objections often raised against the comparative method can now be answered. Critics often point out that the assumption of a unique relationship between adaptation and life style is false. Different species can evolve different adaptations for the same life style: consider the different ways that an antelope and a kangaroo use to move. And a similar form, or behaviour pattern, can evolve for different functions. Territoriality, for example, has evolved more than once, but some animals defend space for mating whereas others do so to keep control over a food resource. A comparative biologist, anxious to test the theory that territoriality is an adaptation for mating, could list many species in support of his theory. However, a truly systematic test would expose his error. But both of these objections apply only to the naive and selective comparative method, which we have already moved beyond.

Even if the comparative method considers all the facts at its disposal, it is still incomplete. There remains the problem of how to count the facts once they have been disposed. My phrase above—"the number of animals in all four categories"—was deliberately vague. For exactly what units are to be counted?

Earlier practitioners (and many simpletons even today) counted species as the unit in their tests. However, scattered through the animal kingdom are groups of closely-related and intensely studied species. They are groups that have been favoured by the whims of biological taste and tradition, or by experimental convenience. The problem is how, in a comparative test, to weigh a large group of closely-related species all of which show similar adaptations, against a small (or less well-studied) group showing different adaptations. I can illustrate this problem by some of my own research. I once studied the relation between the sex of the caring parent in species with uniparental care, and the mode of fertilisation. Species with internal fertilisation (in the female) tend to have maternal care, but species with external fertilisation have paternal care. Most of the externally-fertilising species with paternal care are fish, but I soon ran into the fish family Cichlidae which has many species known to have maternal (mouth brooding) care. Cichlids happen to be an exceptionally well-studied family of a great variety of species. It seemed absurd to weigh all those cichlid species equally with the odd species drawn from 50 or so other families of fish. I did not really know what to do then, but decided to use the family as the unit of my test, so all those cichlids counted as one. Other comparative biologists, working on other ideas, have used species, genera, subfamilies, and complex mixtures of these.

I now prefer a different method. Instead of using only one particular taxonomic level, I use independent evolution as the criterion for counting the frequency of a character. I try to count how many times a character has evolved independently in each environment. The problem with using species, or any other taxonomic level, in the comparative method is that some of the species will share characters by descent from a common ancestor. Not all characteristics will have evolved independently in each species. If a character is counted only as often as it has evolved, then every entry in the test is an independent evolutionary trial of the hypothesis. But although independent evolution may be the correct criterion in theory, there are difficulties in practice. No direct method exists to reveal the habits of ancestral species. Most ancestral species are extinct, and are known only with uncertainty. But uncertainty is not total ignorance (and indeed scientific knowledge is never certain), and it is partially possible to reconstruct ancestors from their descendants. The techniques of cladistic taxonomy were in fact developed to distinguish between characters shared by species because of common ancestry, and characters independently evolved. The usual method is called "outgroup comparison" (see Box).

The cladist uses his method to construct a taxonomy; the comparative adaptationist inverts the technique. He takes a taxonomy for granted, and uses cladism to eliminate taxonomic artefacts. Cladism becomes a method of counting how many times characters (other than those used to construct the original taxonomy) have evolved.

I can show the method in action in a study I have been doing of mating habits. I have been reading the literature on "precopulatory mate guarding" (often abbreviated as "precopula"). In a species with a precopula, the male and female join in close physical contact, but not copulating, for a lengthy time, usually several days, before they eventually do copulate. After copulation the sexes separate, the female to lay her eggs, the male to seek another mate. Many kinds of species have a precopula. It is especially common in crustaceans, such as various kinds of crabs and shrimps. In some kinds of spider, the male and a subadult female "cohabit" for a long time—even all winter—until the female moults to become an adult, upon which they copulate and then separate. Some other, smaller arachnids called mites also move around in a precopulatory embrace before mating. And it is not

Tony Allen

Toads mating, the smaller male grasping the female. He may have been guarding her for several days prior to copulation

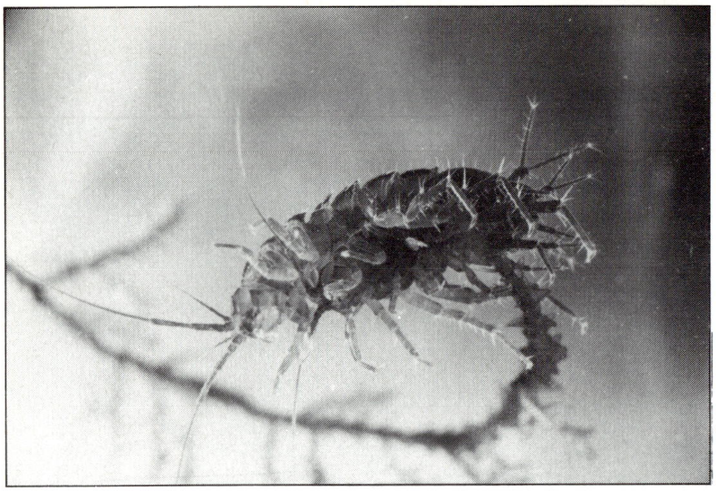

Asellus, the water louse, can copulate only in the 24 hours following each moult. That is why a male guards a female, so that he is in possession when she is able to copulate

uncommon to see pairs of frogs or toads, the smaller male riding on the female, walking to a pond to breed.

The question is whether there is a common factor that has caused all these species to evolve precopulatory mate guarding. In the arthropod groups, the crustaceans and arachnids, the common cause is probably the moult cycle of the female. Arthropods grow by moulting their exoskeletons. In many groups, for a variety of reasons, mating is possible only just after a female has moulted. In spiny lobster and crayfish, for example, the male fixes his spermatophore to the female's exoskeleton. The female loses the spermatophore when she moults, after which a new one is attached. In other groups, such as the freshwater isopod *Asellus* (which looks like a woodlouse and is very common in rivers and lakes in Britain and elsewhere), the female broods her eggs externally and her brood pouch physically prevents copulation. Copulation is possible only during a period of about 24 hours after each moult before the brood pouch has hardened up.

In these arthropods there is a very short period, of a few hours, when the female can mate. So at any one time females that can mate will be rare in the population. If a male could detect, on meeting a female, how long it was going to be until her next moult, it might pay him to stay with her until she moults, if the wait would be short enough. Then the male would be certain to be with the female when mating becomes possible. As males evolved the habit of guarding females before mating, the density of females that could mate would fall even lower, thus increasing the advantage to males that waited with females for some time until the female moult.

The theory predicts that species in which mating can take place only at moulting should have a precopula, but species in which females can mate at any time should not have a precopula. At the species level there is abundant evidence of precopula in species in which mating is at moulting: over one hundred crustacean species mate in this manner. But a cladistic analysis suggests that precopula has evolved only one or two times among the crustaceans, and in the whole animal kingdom about 10 times. In all 10 cases mating is possible only for a short time, and precopulatory mate guarding precedes the copulation itself. There are no unambiguous cases in which precopula has evolved in species in which mating possibilities are not very limited in time. The loss or absence of precopula has evolved about 14 times in the animal kingdom, and in all cases females are fairly continuously available for mating (this is the normal condition in animals). There are again no clear cases of species in which females can mate only at moulting but in which there is no evidence of precopula.

So the comparative evidence strongly confirms the hypothesis. Precopula tends to evolve when females can mate only during a brief period. I have extended the analysis to another habit often associated with precopula: selective mating by males. Another dose of *a priori* reasoning about natural selection will be needed to show why species with precopula tend to evolve into fastidious maters. A male in precopula invests a large amount of time in a particular female. In most of the groups showing precopula, larger females lay more eggs than smaller females. Natural selection, of course, favours males that fertilise more eggs. The Darwinian therefore expects males to prefer to mate with the larger females. Not all the males can mate with large females because they are in limited supply. Now, larger males can usually win fights against smaller males, so the following result might be expected. All the males "want" to mate with large females but only the larger males will succeed. The smaller males will be left with the smaller females. Thus if you were to sample mating pairs from nature you would find homogamy for size: male size and female size would be correlated.

Natural selection does not favour homogamy in species without a lengthy pairing because in these species the larger males can mate with all the females, large and small. Only in species in which mating takes a long time do the smaller males mate at all, and then only with the smaller females.

The comparative method can test this piece of reasoning. I have searched the sparse literature for a connection between homogamy for size and a long duration of pairing (the other criteria, that larger females lay more eggs, and that larger males win fights against smaller males, must also be satisfied). I have uncovered 14 independently evolved cases of homogamy and six cases of random mating which fit the theory. There are, however, six exceptions (three cases of random mating, and three of homogamy, when not predicted). So although the theory that homogamy evolves when mating takes a long time accounts for many cases, it is not the only factor.

My two examples of the comparative method in action demonstrate another point. Critics say that the comparative method can discover only correlations, that it is necessarily silent about evolutionary causality; it cannot tell us which of the correlated factors came first. Take a correlation like that of monogamy and paternal care. Males tend to care for their offspring in monogamous species, but not in polygamous species. Did monogamy evolve first, so making it advantageous for males to help rear the offspring because they have nothing else better to do? Or did paternal care evolve first, so that monogamy would evolve because males lacked the time to search for other mates? In this case, it can be argued in either direction. We do not know which caused which. Now, scientists normally find out about causes by doing experiments. Experiments cannot be done on the causes of large scale patterns in the diversity of life. However, it is sometimes possible to find out about causes simply by thinking about natural selection. In some cases, such as the two examples that I have discussed in detail, natural selection would drive the system in one direction but not in the other.

Consider, for example, the study of homogamy. If mating takes a long time, natural selection will cause the evolution of homogamy. However, if homogamy evolved for some other reason, natural selection would not consequently favour lengthy mating. In this case we know, from thinking about natural selection, the direction of evolution. In a case like the relation between paternal care and monogamy, in which natural selection would drive the system in either direction, it is only reasonable to suppose that evolution has gone in both directions, so there is no single cause.

There are those who maintain that natural selection is an empty tautology, and that the theory of evolution is non-predictive. However, in the comparative method natural selection is used to predict and to explain. It is used to analyse one of the biggest problems in biology. Allied to a sensible method of testing hypotheses, it promises to provide a causal understanding of all the ordered diversity of life.

PART V
BEHAVIOUR AND EVOLUTION

Darwin puzzled over the worker bee's selfless sacrifice in the service of her queen. New theories that focus on the genes rather than the individual carrying them make sense of this behaviour and much more besides. Sociobiology is a useful extension to Darwinism, but there are dangers in sociobiological storytelling

Chapter Thirteen

The games animals play

Animals use rules to settle contests, and the rules change as the stakes change

JEREMY CHERFAS

ETHOLOGISTS have always been concerned with the origins of behaviour. Some devote themselves to discovering how an individual's behaviour is formed, what determines the course of development and to what end. Others worry about long-term influences on behaviour, why one species differs from another and how natural selection operates on behaviour. The two views are not mutually exclusive, and each is important for a full understanding of the other.

Professor John Maynard Smith, of Sussex University, has done a great deal of work on mathematical aspects of evolution and in particular has developed the concept of the evolutionarily stable strategy (ESS) and its application to ethology. An ESS defines the strategy that does best in competition *against itself*. The idea derives from games theory, and specifies what an individual should do in the various "games" of life. For example, in a fight one can be a hawk, always attacking and escalating fights, or a dove, peaceful and retiring. Neither, however, is an ESS because if the population is made entirely of hawks a mutant dove that arises will do better than the hawks, and spread through the population. In the same way a hawk will spread through a population of doves. Maynard Smith showed how to assign costs and benefits to the various strategies (glossing over the problems of finding a common yardstick against which to measure very disparate costs and benefits), and demonstrated that a certain *combination* of hawks and doves exists where the total net gain of the hawks equals the total net gain of the doves. Hawk, then, is as successful as dove. It may be that there are eight hawks to five doves in the population, or that an animal plays hawk eight times to every five he plays dove. The point about an ESS is that it cannot be bettered. Any mutation that arises that does not follow the ESS will not reproduce as well as those that do, and so will not be able to invade the population.

Of course, hawk and dove are not the only strategies in a fight. An animal could be a bully, attacking like a hawk but running away if the attack is returned. This too is not an ESS. Another alternative is to be a dove at first, but fight like a hawk if your opponent is a hawk. This strategy, originally known as Roosevelt (not Franklin D., but Theodore, who said "speak softly and carry a big stick; you will go far"), is called retaliator. It will invade a mixed population of hawks and doves, and eventually oust them to become the ESS.

Another strategy depends on there being some sort of asymmetry between the contestants that is apparent to both of them. If the costs of escalation are high and the benefits of winning low it might pay the animals to use this asymmetry to settle the fight. Maynard Smith calls this strategy bourgeois, and has taken great pains to point out that the asymmetry need have nothing to do with actual fighting ability. Any convention will do to avoid an actual fight.

Nick Davies, when he was at the Edward Grey Institute in Oxford, provided a beautiful example of the bourgeois strategy. Davies studied the speckled wood butterfly, quite plentiful in the Oxfordshire woods. Sun filtering through the trees creates a mosaic of light sunspots on the forest floor, and many of these are occupied by a male speckled wood butterfly. Other males patrol above in the canopy, and occasionally a canopy male flutters down into an occupied territory. The resident flies up to inspect the intruder, there is a brief spiral flight that lasts three or four seconds and then one butterfly carries on up to the canopy while the other flies down to take up residency in the sunspot. Davies marked individual males and discovered that on every single occasion the territory owner won the fight and the intruder returned to the canopy. Perhaps the speckled wood butterfly was using ownership as a bourgeois arbitrary rule for settling disputes.

Ardea

Male speckled wood—a bourgeois butterfly

But there is an alternative explanation. Perhaps in the morning, when the sunspots are first formed, the strongest males are the ones that get a sunspot, and they stay in possession. Davies trapped territory holders and waited until the empty sunspot had been occupied by a canopy male. Then he released the former owner into the sunspot. In all cases the intruder was vanquished, even if the new owner had been on the territory for only a matter of seconds. In another experiment, Davies managed to persuade pairs of males that they were each the owner of a single sunspot; when they eventually met the ensuing struggle escalated and lasted 10 times longer than normal, as if each were convinced that he was in the right.

If Davies is correct, and the speckled woods are bourgeois, we need to know why they have evolved this way. Maynard Smith would predict that the costs of a fight are great and the benefits of winning slight. In other words, it does not pay to fight. But what are the benefits of having a territory? It is not food, because the butterfly follows the sunspot as it tracks across the forest floor, and holds a territory even when the spot falls on the tarmac of the road. So it must be that other great motive force, sex. This intuition is correct, because whereas the territory-holding male gets four courtship flights—with the prospect of mating a female—each day, the canopy males get, on average, only one courtship every five days.

Monarch of the glen, announces his strength with a series of roars

But then why do the males not fight for the increased chance to reproduce? It seems that the answer lies in the relative abundance of sunspots and the behaviour of the residents. There are plenty of sunspots; enough, probably, for all males to have one. And when a male territory holder meets a female they fly up into the canopy to mate, vacating his sunspot to the first comer. So a canopy male has to wait only a little while and he will surely come across an unoccupied territory. In the meantime it would be foolish to risk damage in a fight over a worthless resource; the speckled wood accepts ownership as a bourgeois rule for deciding the outcome of a contest.

Other animals do not have things as easy as the speckled wood butterfly. Tim Clutton-Brock, director of the Large Animal Research Group in Cambridge, has spent over a decade watching the red deer on the island of Rhum. Relevant here are the mating strategies of the red deer stag. In October, the bachelor herds split up and move on to the rutting grounds, where each stag attempts to collect and defend a harem of hinds. An individual stag's reproductive success can be established by counting back from the birth of the calves; there are clear differences between stags, which implies that reproductive strategies have been subjected to strong selection pressure. So just how do "good" males assert themselves? A stag can gain females by successfully challenging other stags. Fights start with a session of roaring, which may be followed by parallel walking and perhaps a tussle with locked antlers. If the challenger had a harem the winner takes control of both sets of hinds. The benefits of a larger harem, for increased reproductive potential, are clear, but what of the costs?

Clutton-Brock estimates that during the average reproductive life of four years, 20 per cent of the stags are permanently injured in fights. The costs are considerable. For a bachelor without a harem the benefits outweigh the costs, and one finds that harem owners initiate fights less often than bachelors. The costs are also increased when both stags hold harems, largely because young stags take advantage of the fight to steal hinds from the harems. (This is known as kleptogyny, but Clutton-Brock and most others prefer "the sneaky fucker strategy".) Fights, and injuries, are more frequent in mid rut, when most conceptions occur.

All these factors point to fights as a prime mover behind increased reproductive success. But in fact there are not that many actual fights and those that do happen are often between stags of the same age and size. Furthermore, stags who initiate fights win more often than would be expected, especially when they have more hinds than the stag they attack. Apparently they know they will win before they approach their adversary. It is as if stags too have a rule for deciding contests, but because the potential benefits accruing to the winner are so large Maynard Smith predicts that the rule will be a "true" one, accurately reflecting the stag's actual fighting ability.

Antler size has little to do with it, nor does weight. Clutton-Brock believes that roaring provides the key. Most attempted intrusions end after the roaring phase, and the rate of roaring is an excellent predictor of the eventual outcome of the match; the winner roars more frequently and longer than the loser. Clutton-Brock's belief that stags do use roaring as a clue to fighting ability was confirmed in a series of playback experiments. At the height of a roaring contest a prime stag can produce about six bellows a minute. Clutton-Brock made up three tape-recordings of roaring, one at 2½, one at 5 and one at 10 roars per minute. He then took a powerful amplifier out to the rutting grounds and played the various tapes to stags. The first, at 2½ roars per minute, did not elicit much response from the resident—after all, a stag that can only manage 2½ roars is hardly worth bothering about. Five per minute was more of a threat, and the stag behaved accordingly, bellowing and threatening towards the loudspeakers and rounding up his hinds. Ten roars per minute is more than any mortal stag could possibly manage. When the harem holder heard this super-stag he "leaped to his feet and looked extremely worried". Because roaring is tiring and difficult to fake, but nevertheless harmless, it makes an admirable substitute for a dangerous fight.

The speckled wood butterfly has little to gain from a fight over territory. Patience will be rewarded, and accidental intrusions onto another's sunspot are quickly settled with no recourse to fights or even substitute fights. The red deer stag, on the other hand, has much to gain and much to lose. He cannot afford to be as bourgeois as the butterfly, but nor can he afford to risk unnecessary injury. His strategy is to use roaring as a predictor of his opponent's fighting ability before engaging in a real fight. These two studies show that, with the right approach, we can make sense of animal societies.

Maynard Smith's game theory attack on animal behaviour is starting to pay dividends. We now begin to understand the forces behind animal competitions. Some sceptics have been worried by the apparent simplicity of the models and find it difficult to believe that this simple approach can throw any light on the complicated business of real life. Maynard Smith has the last laugh: "When I was an aircraft designer," he once told me, "we built our planes on the assumption that air was incompressible. We all knew that air was *not* incompressible, but the planes flew nevertheless." Game theory may be simple, but it works.

Chapter Fourteen

Genes take care of their own

Sometimes animals selflessly help their fellows, but in fact they are helping themselves

JEREMY CHERFAS

BIOLOGY is replete with examples of bizarre phenomena. Take the selfless sacrifice of the worker honey-bee in the service of her queen. The worker is sterile, and cannot herself reproduce, yet she toils to nurture the offspring of another bee. She is even willing to die in defence of the hive. How can one account for this altruistic behaviour that so obviously goes against the fundamental principle of life—reproduction is everything. One might speculate from the comfort of one's armchair that the worker makes her sacrifice on the altar of the species. She does what she does because it enables honey-bees to survive.

But think what would happen if a mutant bee, one who cheated the system, were to arise. Such a bee might be resistant to the substances the queen secretes that keep her workers sterile, and might herself reproduce, allowing others to care for her young. A cheat bee might also hang back when the hive is attacked, secure in the knowledge that other suckers will defend her. This cowardly reproductive bee would have cowardly reproductive offspring, and in no time the selfless workers would be outnumbered by cheats. Who then would care for and defend the hive? And what would happen to the species if no bees were prepared to raise the young and the hives were always open to attack? They would no longer be social honey-bees, though some lines might survive as solitary bees.

John Mason/Ardea

A queen bee (marked with a blob) attended by her daughters

A "good of the species" argument is intuitively appealing, but nevertheless it is clearly impossible. New developments in biology have led to real understanding of altruism in general, not just as it applies to bees, and along the way have suggested solutions to many other knotty problems and revitalised the whole field.

The keystone of Darwin's account of evolution by natural selection was the idea of fitness. In every generation there are those who are better adapted to deal with the problems posed by life. If the environment is becoming colder, animals with a thick coat will be better adapted than animals with a thin coat. More of the thick-coated animals will survive and breed, and if thickness of coat is determined by genes the proportion of thick-coated animals in the population goes up. Fitness, crudely, is the ability to leave offspring that will in turn survive to breed, and any change that increases fitness will be selected.

Worker bees are clearly not fit, for they leave no offspring at all. How could evolution ever produce an animal that is sterile? Inclusive fitness, an idea that owes something to J. B. S. Haldane but was fully developed by Bill Hamilton, provides the answer: where Darwinian fitness concentrates on the individual, inclusive fitness concentrates on the individual's extended family. To understand inclusive fitness we need to know a little about reproduction and genetics.

The so-called genetic blueprint is contained in the chromosomes which, with the exception of the sex chromosomes, come in pairs. Man has 23 pairs of chromosomes, the fruit fly only 8. Genes exist in alternative forms, called alleles, and an individual normally has two alleles for any given gene, one on each chromosome of a pair. (This is not, as it happens, the case with bees, where the males have only one chromosome of each pair while the females have both, and although this changes the mathematics of what follows it does not change the conclusions.) Thus, to take one of Gregor Mendel's famous examples, there is a gene for seed-coat texture in the pea and it has two alleles, smooth and wrinkled. A pea can have two smooth alleles, or two wrinkled alleles, or one of each. In reproduction the offspring receives one chromosome of each pair from each parent, and it is a matter of chance as to which of a parent's two alleles for each gene the offspring gets. Thus, from the point of any single parental allele there is a one in two chance that it will be passed on to the offspring. This probability is the same for all the genes in all the chromosomes, so that the relatedness (Hamilton's term) of an offspring to either one of its parents is one in two, or a half.

All offspring of the same pair of parents are related to each parent by one half, but how are they related to each other? Well, looking again at just one gene for the sake of simplicity, there is a one in two chance that a given allele will be passed on to offspring *A*, and a similar chance that the same allele will be given to offspring *B*, so the chance that it is present in both *A* and *B* is one in four. If *A* and *B* share one parent they will be related to each other by one quarter, as there is a one in four chance that they both have the same copy of any particular gene. If they share both parents their relatedness is one half, because they are related by one quarter twice. This sort of quantification can be carried out for all sorts of relationships. Grandchildren, for example, are related to each of their grandparents by one quarter, and cousins by one eighth.

All these sums are fine, but how do they help us to understand evolution? Well, the relatedness of two individuals is the probability that they share a particular allele, and the purpose, if you will, of each allele is to maximise the number of copies of it that survive. Two of your full siblings are likely between them to contain a copy of any of your alleles, so it matters not *to the allele* whether you or your two sibs survive to breed. It is this view of relatedness that led to Haldane's celebrated pronouncement, made in a pub after some hasty scribbles on the back of an envelope, that he would lay down his life for two of his brothers or eight of his cousins.

Inclusive fitness, then, is the total fitness of all your relations. devalued appropriately by their relatedness to you. If you help two of your sibs to have 10 offspring each it is, in terms of inclusive fitness, exactly the same as having 10 offspring of your own. The concept is not restricted to "relations" as we normally think of them, but the mathematics get harder as the blood ties get weaker, and the best examples of inclusive fitness in action do refer to quite close relations. In such a case the selection process is called kin selection, to distinguish it from Darwinian individual selection.

The worker bee is helping to rear her own sisters. In forgoing reproduction and assisting her mother she maximises her inclusive fitness, and ensures the survival of many more copies of her genes than she could on her own. Inclusive fitness helps us to understand the evolution of sterile worker castes in insects, where the family ties are rather clear (see Box). It also provides a powerful analytical tool for taking apart other societies, especially those that include a lot of behaviour that looks to be altruistic.

A very extensive group of relatives

A pride of lions provides a perfect example of a society that apparently runs on cooperation and altruism. The females form the stable core of the pride. They hunt together, and are able to capture large animals sufficiently often to sustain the whole pride, including sick lions who make no contribution to getting food. While some females are out hunting, others stay behind with the cubs, suckling all and sundry rather than solely their own offspring. These babysitters will also feed from the kill, as will the males, who do not do much of anything. From time to time male rule over the pride is challenged by young bloods from other prides, and occasionally the challenge is successful. The old males are driven out, often to die, and the new males take over. When this happens the incoming males often kill the cubs in the pride.

The lion pride defies comprehension in simple evolutionary terms, but thanks to the painstaking work of Brian Bertram in the Serengeti National Park we now have some understanding of how the pride works. The problem in conveying this understanding lies in knowing where to break into the life cycle of the lion. I will start with the arrival of new males.

Behind the successful takeover of a pride will be a combination of factors. The more challengers there are, the more likely it is that they will succeed. Success is also more likely if the resident males are old and few in number. The new lions often kill off the cubs. This is understandable, as the cubs are the offspring of the previous incumbents, and their continued survival will only delay the start of the new males' reproductive effort. Lions hold a pride only once; it is their sole chance to breed and any behaviour that allows them more breeding during their short tenure will be selected. What is not so understandable is why the females do not put up much of a fight when their young are being massacred. After all, the cubs represent an investment for the females, and though they could not fight the males directly, one can imagine a different kind of pride in which the lionesses would help the lions to withstand a challenge if there were young cubs in the pride. The reason the females do not behave like this will become apparent.

Ardea

A pride of lions. As Brian Bertram says, "the family that preys together, stays together"

New lions in a pride are also able to get rid of cubs as yet unborn; after a takeover those lionesses that are pregnant often abort. This is akin to the Bruce effect that was first demonstrated in laboratory mice, and is probably due to chemicals in the urine of the new males.

Why bees, ants and wasps are so often social

THE Hymenoptera—bees, ants and wasps—contain a great many social species. The workers are genetically female, but produce no offspring themselves. Instead, they dedicate themselves to helping their mother, the queen, to produce more offspring. They care for the eggs and young, guard the nest and take care of the queen. Social behaviour has evolved independently about 11 times in the Hymenoptera, and always the sterile workers are females. Kin selection explains why the workers are females, and why there is a tendency to evolve sociality. The explanation hinges on the curious genetics of the Hymenoptera.

The queen has the choice of whether to fertilise an egg or not. Females develop from a fertilised egg, and so are diploid. Males, by contrast, develop from an unfertilised egg and are haploid; they contain but a single chromosome of each pair. All offspring get half the queen's genes, so they are all related to their mother by a half. But sisters are more closely related to one another than they are to their mother. This is because the half set of their genes that they get from their father are all identical. So through their father alone the females are totally related. The rest of their genes they get from their mother, and there is a one in two chance that a particular maternal gene will be present in any given pair of sisters. The total relatedness is the average of one—via the father—and a half—via the mother: three-quarters. So it could pay them to forgo their own reproduction (producing their own offspring to whom they would be related by a half) in favour of helping their mother raise more sisters, to whom they are related by three-quarters. Males are equally related to offspring and siblings so this kin selection pressure for them to become altruistic helpers does not exist.

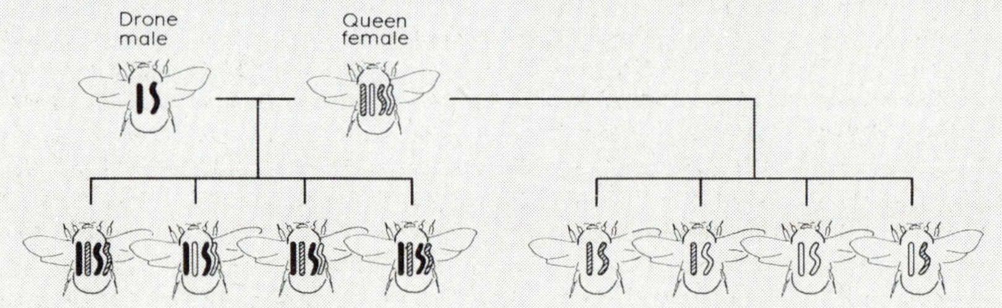

The death of the cubs, born and unborn, means that the lionesses quickly become fertile and sexually receptive. Moreover, they all tend to come into oestrus at more or less the same time. They copulate with the new males and eventually give birth to their cubs, again more or less synchronously. We now have several cubs and several lactating mothers in the pride. A lioness will give milk to cubs other than her own, so the more females there are with milk the more likely it is that a cub will be able to feed while its mother is away on a hunt. For this and other reasons, cubs born out of synch often do not last long. Females can benefit from the death of their previous cubs if it means their present cubs have a better chance to survive.

When the lionesses make a kill, and it is they who make most of the kills, all members of the pride will get to eat, as long as times are good and kills frequent and large. When times are not so good the lions come along and bully their way to the food. If there is any left the lionesses get it, and only then do the cubs get a look in. This is tough on the cubs, who often starve, but it ensures that the lionesses get enough food to sustain the next hunt. Besides, when supplies improve it is easier to produce a cub, with its short period of gestation, than a full grown animal capable of killing the improved supplies.

If times are good the cubs will grow and mature. What happens to them next depends on whether they are male or female. Females will be accepted into the pride if there is room, but otherwise will be driven out. If they are very lucky they may find an unoccupied space in which to set up a new pride, but by and large these outcast females never breed. Males are invariably driven out of the pride. A cohort born together will wander about looking for a pride to take over, closing the circle.

Recall that a larger group of males stands a better chance of getting a pride and keeping it. They will father more cubs as a result. Recall too that males who kill cubs engender synchrony of births in the females, and that cubs born in synchrony are more likely to survive. The more cubs that survive, the larger will be the group of males driven out to find a new pride. They will be more successful and so cub-killing will spread through the population.

Communal suckling would spread for similar reasons. Perhaps at first a lioness's daughters would stay with her in the pride, so that in suckling other cubs the lioness would be caring for her grandchildren. Cubs survive better in larger groups, so as she suckles other cubs the lioness ensures that her own offspring grow up with companions. Her sons will hold on to prides longer and her daughters will be more successful because they too will suckle other cubs. Infanticide on the part of incoming males synchronises the females, and makes it easier for communal suckling to occur. Put the two tendencies—males who kill cubs that are not their own and females who suckle cubs that are not their own—together, and you have a cooperative social system that will spread through a population and, once established, maintain itself. And if the females that started the crèche were related to each other, as sisters, or mother and daughters, they would be helping their relatives, and the system would spread even more quickly.

Bertram watched four prides for several years, and built up a composite picture of the typical pride. From such a pride he calculated that the males were more or less half brothers (related by about one quarter) while the females were approximately full cousins (related by one eighth).

Brian Bertram's work shows how the various threads of lion life can be unravelled, and inclusive fitness provides a way to see how each thread contributes to the fabric. By concentrating on the separate bits of behaviour, each of which appears at first sight to decrease the individual fitness of the animal concerned, we see how they fit together in a way that in fact increases the likelihood that they will survive into the next generation.

Not all societies are lion prides, but all are founded on similar principles. Among monkeys, there are many species in which males take possession of a group of females, but only a few in which the new males kill the infants. A wolf pack contains related females, but only one of them breeds. Hunting dogs cooperate to bring down large prey, but even when times are hard it is the pups that have priority at the kill. Slowly the similarities and differences between various societies are beginning to make sense, and they do so in the light of inclusive fitness.

The secret of our new understanding of behaviour, especially social behaviour, is to forget about the individual and think about the gene. Copies of an individual's genes are likely to be present not only in its own offspring but also in its close relatives and their offspring. Any gene that strengthens the tendency to help relatives will assist in its own spread and survival. The tendency to help relatives seems to be at the bottom of all societies. And at the bottom of perhaps the greatest revolution in biology since 1859 is Bill Hamilton's notion of inclusive fitness.

Chapter Fifteen

Sex around the cow-pats

Careful observation of dungflies as they mate on and around cow-pats reveals that they use sophisticated strategies in maximising their reproductive success

GEOFFREY PARKER

WHY do peacocks sport outrageously resplendent plumage compared with their more conservative mates? And why do majestic red deer stags engage in ferocious combat with each other over possession of harems, risking severe injury from their spear-point antlers? Charles Darwin described these and many other similar phenomena as the products of sexual selection—any inherited characteristic that helps an individual produce more offspring than its competitors will become common in the population. Sexual selection, Darwin argued, operates in two ways: first, females may exert an active choice of their mate, forcing males to display more and more attractive features (the peacock is a good example); and secondly, the males may compete directly with each other, in which case they will develop effective means of doing so (the stags' antlers and fighting ability is an example here). In either case, the physical and behavioural characteristics are evolutionary responses to males competing for females.

Why the competition? If males and females invested expensively and equally in their offspring, there would be little advantage to be gained in individuals of one sex attempting to monopolise the sexual favours of the other—no individual would have the spare resources with which to do it. But for the most part this is not so. For a start, female sex cells (eggs) are usually much bigger—and therefore much more costly to produce—than male sex cells (sperm). And, as exemplified by mammals, females may be involved in a good deal of expenditure in care of the offspring before and after they are independent. Returning once again to the example of red deer, we can see that a male, whose only involvement in its offspring is to deposit sperm into a female when she is in oestrus, has everything to gain from competing with its fellow males in order to fertilise as many females as it can. A hind's interest in sexual encounters, by contrast, terminates when she has mated just once—she is fertilised and has no more to gain.

Darwin, and more recently A. J. Bateman (some 30 years ago), first alluded to the notion of relative parental investment as the basis of sexual selection. Since then Robert Trivers, while at Harvard, developed the concept further and suggested that the intensity of competition depends on the degree of disparity in parental investment. Mostly, but not exclusively, it is the males who invest less in their offspring, and therefore evolve physical and behavioural characteristics to aid in competition with each other. The evolution of such characteristics is of course by the passive process of natural

Dungflies grapple for a female, the successful male mates, and stays mounted while the female lays her eggs

selection; sexual selection is simply a particular type of natural selection.

As I have indicated, nature abounds in examples of sexual selection—the peacocks' feathers and the stags' behaviour are *qualitative* support for the theory of sexual selection. However, more detailed *quantitative* tests of sexual selection theory are necessary if we are to assess its importance as a force in evolution. I therefore decided to study the mating behaviour of dungflies (*Scatophaga stercoraria*). My approach was to make predictive models of optimal mating strategies. Then, notebook, stopwatch and camera in hand, I set out among the cow-pats to discover what the flies really did.

Spring and autumn days bring large numbers of dungflies to cow-pats, especially freshly dropped ones. The golden-yellow males, which are slightly bigger than the females, often arrive at a pat within seconds after it hits the ground. They appear to locate their target by smell, often flying long distances from downwind. The females, heavy with eggs, arrive more slowly and usually overshoot the dung, landing a short distance upwind. They then turn and walk back to the pat, encountering males on the way, or when they reach their destination. Females are fertilised around droppings and lay their eggs in the dung.

Males, who usually outnumber females by four or five to one at the pat, search for potential mates both on the dropping or in the upwind grass. When a male spots a lone female he leaps, mounts, and mates immediately—there is little hint of courtship. Then, when the male has finished depositing his sperm, instead of dismounting and taking his leave, he simply disengages genital contact and stays put until the female has finished laying her eggs in the dung—a rather unusual behaviour. The whole process of encounter, mating and egg-laying takes on average 156.5 minutes.

When the female has laid her batch of eggs (usually about 43), she sways from side to side a few times, the male dismounts, and she flies quickly away. If she survives she visits another pat a week or so later when she has more eggs ready to be laid; in the meantime she avoids fresh dung and harassment by males. (Incidentally, when a female mates she receives sperm sufficient to fertilise four batches of eggs; the fact that she mates each time she visits a pat in order to lay the eggs is therefore interesting.)

The reason for the male's persistence after he has copulated with the female becomes clear after a short spell of observation at a fresh pat. The males consider all females as fair game, whether or not they are already mounted. Mounted females and their attendant mates are repeatedly attacked by single males. Although the male deploys an elaborate series of defensive manoeuvres designed to keep the attacker from the female, he is not always successful, especially if two males close in at the same time. If the attacker manages to obtain a grip on the egg-laying female a prolonged struggle ensues. And if the incumbent is displaced, the usurper mates with the female before allowing her to finish laying her eggs. The chances of a male being dislodged from an egg-laying female are about 1 in 60; for a copulating pair intrusions are successful at half this rate, about 1 in 150.

It seemed to me that the behaviour of male dungflies, with their intense competition for females, offers impressive qualitative evidence for sexual selection. My aim, however, was to make quantitative assessments, and in my calculations I assumed that selection would favour strategies that maximise a male's rate of fertilisation throughout his reproductive life. I planned to calculate, for instance, how long a male should try his luck at a particular pat, how long he should remain copulating with a female and where his best chances of encountering a female would be, and then compare this with real life. A match between calculation and the flies' actual behaviour would be strong support for the theory of sexual selection.

A field full of cows provides many potential pats where females will want to lay their eggs. And as females prefer

A: How long to wait

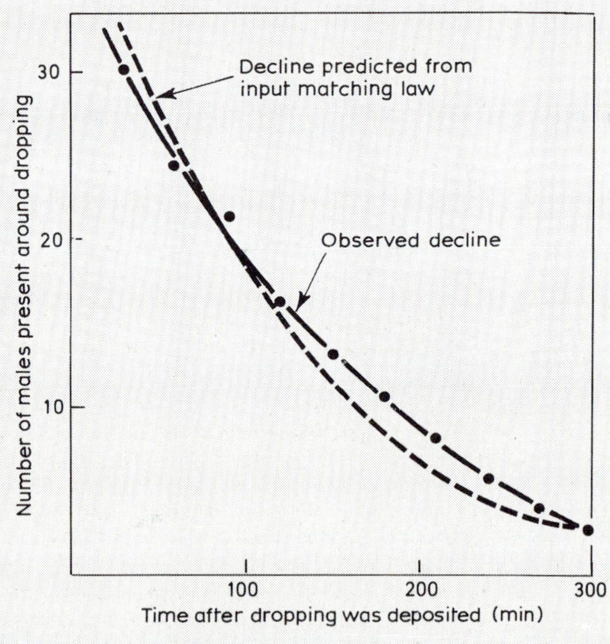

The graph shows the calculated and observed rates of decline in the number of males at a dropping as it gets older. The evolutionarily stable strategy (ESS) for the stay-time is calculated as follows: males should arrive as quickly as possible after the pat drops and all should stay initially; after a certain critical time males should leave according to an input matching law. This law states that the number present should equal some constant multiplied by the rate at which females arrive. The constant is the mean (averaged throughout all reproductive time) rate of capture of females by males. Observation accords with calculation, implying that each male achieves the same rate of fertilisation even though no two males stay exactly the same length of time.

fresh pats to stale ones (their rate of arrival declines exponentially as the pat gets older), males might seem to do well by flitting from one fresh dropping to another. A disadvantage of this mobile strategy, however, is that the longer a male spends in travelling around the field, the less time he will have to search for females around the fresh droppings. A further complication is the behaviour of competitors: if a dropping is swarming with other males it is clearly better to move on to a less densely populated area and *vice versa*.

The balance, then, is between the pat's freshness and the number of females arriving (which are directly related) and the number of competitors. Working with Bob Stuart, I calculated that the so-called evolutionarily stable strategy for the males is to arrive as quickly as possible after the pat is dropped and for all males to stay initially; after a certain critical time the males should leave according to a mathematical law involving the rate at which females arrive (see Box **A** for a more detailed explanation). We drew a graph which predicted how many males should be present at a pat, given the factors I have mentioned, if all males are to have an equal chance of fertilising a female. As Box **A** shows, the match between prediction and observed behaviour is very close. Although we can say that the flies' behaviour in fixing on the length of time they should stay at a pat is, in evolutionary terms, stable, we do not know (genetically and physiologically) how they achieve it.

Why do males guard their females while they are laying eggs? Now, in dungflies, as in many insects, the eggs are fertilised as they pass down the egg-laying tube. This means that even if a male has deposited sperm in a female he cannot be certain that the eggs will have been fertilised by his sperm until they are safely laid in the dung. If a second male copulates with a female just before or while she is laying eggs, the sperm of the first male are displaced—with a full batch of eggs, 80 per cent will be fertilised by sperm from the second male and only 20 per cent by those of the first male. A male who forsakes his female immediately after copulating with her therefore risks losing three-quarters of his potential offspring to an interloper. An individual who, genetically, pursues this strategy is destined for evolutionary oblivion. So, even though guarding a female while she lays her eggs costs the male some 17 minutes—during which time he could be searching for new mates—the strategy is well worthwhile.

Before allowing a female to lay her eggs, the male has to "decide" how long he should spend copulating with her: given that the female already contains sperm from previous matings, the more sperm a male inserts during copulation, the more eggs will be fertilised by his sperm rather than by that of other males. But, as I determined in laboratory experiments, the benefits of prolonging copulation are subject to exponential diminishing returns (Box **B**). Moreover, a male that wastes time in long copulations is missing opportunities for getting on with the next encounter (after standing guard during the egg-laying process of course). Natural selection should therefore *optimise* sperm displacement through copulation rather than *maximise* it.

Our calculation for the optimum time for copulation worked out to be 41 minutes (see Box **B** for detailed explanation). In fact, the average turned out to be slightly shorter: 36 minutes. The discrepancy may be due to the fact that in our mathematical model we assumed that "extra" sperm "cost" nothing, whereas they may well have a biological cost that is similar to a time penalty. Nevertheless, the match between prediction and fact is encouraging and shows we are on the right track.

Males occasionally capture females just as they are leaving the pat after having laid their eggs. Although mating with such a female will probably secure some offspring, the number will be greatly reduced because she will certainly mate once again before laying her eggs in a week's time. There is of course the possibility that the female will not survive to lay again, in which case the sperm will have been totally wasted. On average, mating with a female that has just laid her eggs is worth about 3 offspring as against 35 for a full female (some of whose eggs will be fertilised by other males). Mathematically, we predict that males should mate for just 18 minutes with "empty" females (as against 36 minutes for a full one). In fact, the males spend 15 minutes with these females. Once again, it seems that the males are adapted to optimise their behaviour in this special circumstance.

Males also face the problem of where to mate: as the male can fly the female to where he wants, he has the choice of taking her to the pat, upwind or downwind. The pat seems a good bet as this is where the female lays her eggs. What is more, the extra heat of the fresh dung speeds up copulation

B: How long to couple

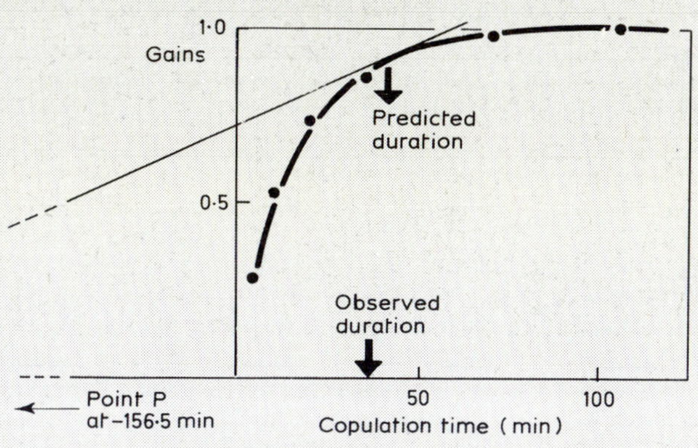

Calculation of the optimal time for which a male should copulate with a female. Imagine that the x axis (copulation time) is extended beyond the origin to a point P equivalent to the mean time taken to find a female and guard her while she lays her eggs. (This is 156.5 minutes, from our field observations.) The optimum mating time is now given by the tangent to the gain curve drawn from point P, giving an answer of 41 minutes, five minutes longer than that observed.

C: Where to search

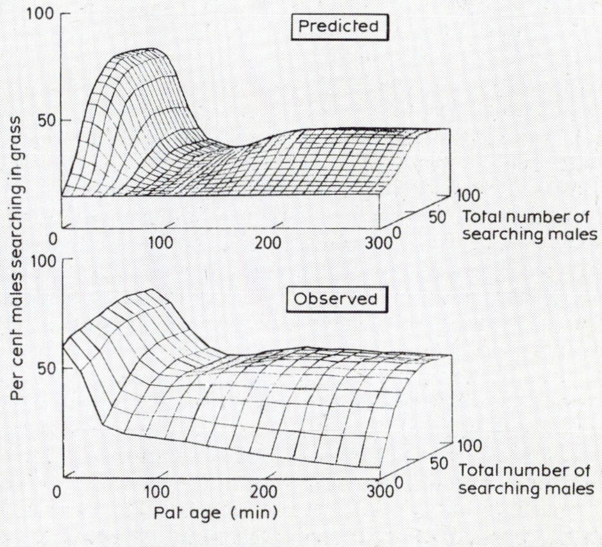

Above: predicted profile for male search strategy plotted against age of the pat and the total number of searching males, assuming that rates of fertilisation achieved by searching on the dung are equal to those achieved by searching in the grass. Below: observed profile for male search strategy. The fit between the two is encouraging.

and this can give a male a bonus few extra minutes which he can use later in new searches for females. The snag is that the pat is the ideal place for intruders to operate. I calculated that if there were less than five other males at the dropping the benefits of staying outweighed the costs. More than five, and departure was the best predicted strategy. Happily, these numbers were confirmed when we looked at real life (Box **C**).

If staying at the pat becomes too dangerous, where is the safest evacuation site, downwind or upwind? As the chances of being displaced are about the same upwind as they are on the dropping, it seems sensible to take refuge downwind. Again, this is borne out by observation. The reason that single males congregate in the upwind grass (and on the pat) rather than downwind is that there is a better chance of meeting single incoming females. The benefits of waiting in the downwind area, where incoming females are rare, hoping for possible opportunities of displacing a mating competitor, are clearly small. In my observations I also noticed that a male who meets a female in the upwind grass almost always flies her downwind (if the pat and upwind grass is dangerously overpopulated) where they then mate.

Finally, what is the best place in which males should wait and search for incoming females—on the pat or in the upwind grass? If there were no competitors in the area, a male's best strategy would obviously be to wait on the pat until the females arrive. Individuals are normally not so lucky, but until there is a significant number of other males present, waiting on the dropping gives the best payoff. What is more, takeovers of egg-laying females also offer a reasonable prospect (giving a mean value of 18 eggs fertilised). When the competition becomes stiff, however, a move upwind is called for. As more and more males arrive they should, according to the sexual selection theory, distribute themselves so that each individual maximises its chances of fertilisation. As a result there should be an equilibrium distribution of males over the pat and the upwind grass, depending on the number of competitors present and the rate at which new females arrive. Because competition is tight, each male will have roughly the same payoff, whether it is on the pat or in the grass.

I calculated equilibrium distribution patterns for such conditions, and discovered in my observations that the flies behave very much as predicted in the model. This, together with the previous successful predictions I've mentioned, encourages me to believe that sexual selection is a powerful evolutionary force and that it can act with more precision than has been generally imagined. Complex patterns of behaviour can be evoked in response to quantitative measurement of several clues in the environment. I should stress once again, however, that the behaviour patterns evolve through passive selection of the most fruitful strategies. Evolutionary biologists often talk about natural selection as though it were in some way purposive, but this is only a convenient shorthand. Individuals vary, and some of the variations are more successful than others. These spread through the population, and that is all there is to natural selection. We often talk in terms of purpose, but variation and selection know no purpose.

Chapter Sixteen

Sociobiology: the art of storytelling

Too many sociobiological explanations of behaviour come into the category of "Just So" stories. They may be plausible but are less than rigorously supported by solid evidence

STEPHEN JAY GOULD

LUDWIG VON BERTALANFFY, a founder of general systems theory and a holdout against the neodarwinian tide, often argued that natural selection must fail as a comprehensive theory because it explains *too* much—a paradoxical, but perceptive, statement. In 1969 he wrote: "If selection is taken as an axiomatic and *a priori* principle, it is always possible to imagine auxiliary hypotheses—unproved and by nature unprovable—to make it work in any special case . . . Some adaptive value . . . can always be construed or imagined.

"I think the fact that a theory so vague, so insufficiently verifiable and so far from the criteria otherwise applied in 'hard' science, has become a dogma, can only be explained on sociological grounds. Society and science have been so steeped in the ideas of mechanism, utilitarianism, and the economic concept of free competition, that instead of God, Selection was enthroned as ultimate reality."

Similarly, the arguments of Christian fundamentalism used to frustrate me until I realised that there are, in principle, no counter cases and that, on this ground alone, the theory is bankrupt.

The theory of natural selection is, fortunately, in much better straits. It could be invalidated as a general cause of evolutionary change. (If, for example, Lamarckian inheritance were true and general, then adaptation would arise so rapidly in the Lamarckian mode that natural selection would be powerless to create and would operate only to eliminate.) Moreover, its action and efficacy have been demonstrated experimentally by 65 years of manipulation within *Drosophila* bottles—not to mention several thousand years of success by plant and animal breeders (see chapter twelve).

Yet, in one area, unfortunately a very large part of evolutionary theory and practice, natural selection has operated like the fundamentalist's God—he who maketh all things. Rudyard Kipling asked how the leopard got its spots, the rhino its wrinkled skin. He called his answers "just-so stories". When evolutionists study individual adaptations, when they try to explain form and behaviour by reconstructing

How the rhinoceros got his skin

National Trust

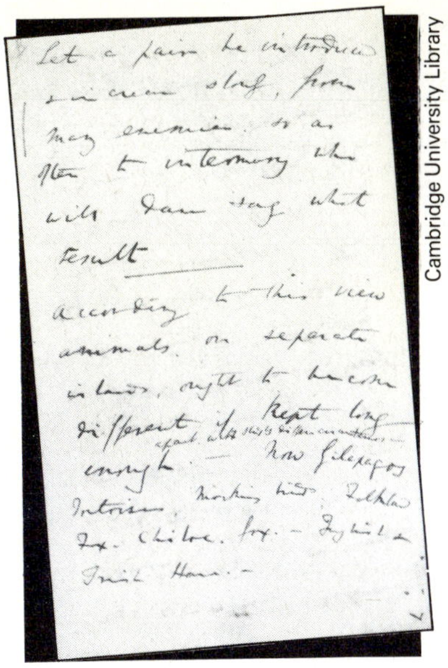
Cambridge University Library

A page from Darwin's notebook

history and assessing current utility, they also tell just-so stories—and the agent is natural selection. Virtuosity in invention replaces testability as the criterion for acceptance. This is the procedure that inspired von Bertalanffy's complaint. It is also the procedure that has given evolutionary biology a bad name among many experimental scientists in other disciplines. We should heed their disquiet, not dismiss it with a claim that they understand neither natural selection nor the special procedures of historical science.

This style of storytelling might yield acceptable answers if we could be sure of two things: first, that all bits of morphology and behaviour arise as direct results of natural selection, and secondly, that only one selective explanation exists for each bit. But, as Darwin insisted vociferously, and contrary to the mythology about him, there is much more to evolution than natural selection. Darwin was a consistent pluralist who viewed natural selection as the most important agent of evolutionary change, but who accepted a range of other agents and specified the conditions of their presumed effectiveness. In chapter seven of *The Origin of Species* (sixth edition), for example, he attributed the cryptic coloration of a flatfish's upper surface to natural selection and the migration of its eyes to inheritance of acquired characters. He continually insisted that he wrote his two-volume *Variation of Animals and Plants Under Domestication* (1868), with its Lamarckian hypothesis of pangenesis, primarily to illustrate the effect of evolutionary factors other than natural selection. In a letter to *Nature* in 1880, he used the sharpest and most waspish language of his life to castigate Sir Wyville Thompson for caricaturing his theory by ascribing all evolutionary change to natural selection.

Since all theories cite God in their support, and since Darwin comes close to this status among evolutionary biologists, the panselectionists of the modern synthesis tended to remake Darwin in their image. But we now reject this rigid version of natural selection and grant a major role to other evolutionary agents (genetic drift, fixation of neutral mutations, for example). We must also recognise that many features arise indirectly as developmental consequences of other features directly subject to natural selection. Moreover, and perhaps most importantly, there are a multitude of potential selective explanations for each feature. There is no such thing in nature as a self-evident and unambiguous story.

When we examine the history of favoured stories for any particular adaptation, we do not trace a tale of increasing truth as one story replaces the last, but rather a chronicle of shifting fads and fashions. When Newtonian mechanical explanations were riding high, G. G. Simpson wrote (in 1961): "The problem of the pelycosaur dorsal fin . . . seems essentially solved by Romer's demonstration that the regression relationship of fin area to body volume is appropriate to the functioning of the fin as a temperature regulating mechanism". Simpson's firmness seems almost amusing since now—a mere 20 years later with behavioural stories in vogue—most palaeontologists feel equally sure that the sail was primarily a device for sexual display. (Yes, I know the litany: it might have performed both functions. But this too is a story.)

On the other side of the same shift in fashion, a recent article on functional endothermy in some large beetles had this to say about the why of it all: "It is possible that the increased power and speed of terrestrial locomotion associated with a modest elevation of body temperatures may offer reproductive advantages by increasing the effectiveness of intraspecific aggressive behaviour, particularly between males". This conjecture reflects no evidence drawn from the beetles themselves, only the current fashion in selective stories. We may be confident that the same data, collected 20 years ago, would have inspired a speculation about improved design and mechanical advantage.

Most work in sociobiology has been done in the mode of adaptive storytelling based on the optimising character and pervasive power of natural selection. As such, its weaknesses of methodology are those that have plagued so much of evolutionary theory for more than a century. Sociobiologists have anchored their stories in the basic Darwinian notion of selection as individual reproductive success.

Sociobiologists have broadened their range of selective stories by invoking concepts of inclusive fitness and kin selection to solve (successfully I think) the vexatious problem of altruism—previously the greatest stumbling block to a Darwinian theory of social behaviour. (Altruistic acts are the cement of stable societies. Until we could explain apparent acts of self-sacrifice as potentially beneficial to the genetic fitness of sacrificers themselves—propagation of genes through enhanced survival of kin, for example—the prevalence of altruism blocked any Darwinian theory of social behaviour (see chapter fourteen).)

Thus, kin selection has broadened the range of permissible stories, but it has not alleviated any methodological difficulties in the process of storytelling itself. Von Bertalanffy's objections still apply, if anything with greater force, because behaviour is generally more plastic and more difficult to specify and homologise than morphology. Sociobiologists are still telling speculative stories, still hitching without evidence to one potential star among many, still using mere consistency with natural selection as a criterion of acceptance.

Kenneth Fink/Ardea

Mountain bluebird. Can we really assume that the male permits adultery once he has fertilised the female?

David Barash, for example, tells the following story about mountain bluebirds. (It is, by the way, a perfectly plausible story that may well be true. I only wish to criticise its assertion without evidence or test, using consistency with natural selection as the sole criterion for useful speculation.) He reasoned that a male bird might be more sensitive to intrusion of other males before eggs are laid than after (when he can be certain that his genes are inside). So Barash studied two nests, making three observations at 10-day intervals, the first before the eggs were laid, the last two after. For each period of observation, he mounted a stuffed male near the nest while the male occupant was out foraging. When the male returned he counted aggressive encounters with both model and female. At time one, males in both nests were quite aggressive towards the model and less, but still substantially, aggressive towards the female as well. At time two, after eggs had been laid, males were less aggressive to models and scarcely aggressive to females at all. At time three, males were still less aggressive towards models, and not aggressive at all towards females.

Barash concludes that he has established consistency with natural selection and need do no more. "These results are consistent with the expectations of evolutionary theory. Thus aggression toward an intruding male (the model) would clearly be especially advantageous early in the breeding season, when territories and nests are normally defended . . . The initial, aggressive response to the mated female is also adaptive in that, given a situation suggesting a high probability of adultery (that is, the presence of the model near the female) and assuming that replacement females are available, obtaining a new mate would enhance the fitness of males . . . The decline in male-female aggressiveness during incubation and fledgling stages could be attributed to the impossibility of being cuckolded after the eggs have been laid . . . The results are consistent with an evolutionary interpretation. In addition, the term 'adultery' is unblushingly employed in this letter without quotation marks, as I believe it reflects a true analogy to the human concept, in the sense of Lorenz. It may also be prophesied that continued application of a similar evolutionary approach will eventually shed considerable light on various human foibles as well."

Consistent, yes. But what about the obvious alternative, dismissed without test in a line by Barash: male returns at times two and three, approaches the model a few times, encounters no reaction, mutters to himself the avian equivalent of "it's that damned stuffed bird again", and ceases to bother. And why not the evident test: expose a male to the model for the *first* time *after* the eggs are laid.

We have been deluged in recent years with sociobiological stories. Some, like Barash's, are plausible, if unsupported. For many others, I can only confess my intuition of extreme unlikeliness, to say the least—for adaptive and genetic arguments about why fellatio and cunnilingus are more common among the upper classes, or why male panhandlers are more successful with females and people who are eating than with males and people who are not eating.

Not all sociobiology proceeds in the mode of storytelling for individual cases. It rests on firmer methodological ground when it seeks broad correlations across taxonomic lines, as between reproductive strategy and distribution of resources, for example, or when it can make testable, quantitative predictions as in Bob Trivers and Hope Hare's work on haplodiploidy and eusociality in Hymenoptera. Here sociobiology has had and will continue to have success. And here I wish it well. For it represents an extension of basic Darwinism to a realm where it should apply.

Sociobiological explanations of human behaviour encounter two special difficulties, suggesting that a Darwinian model may be generally inapplicable in this case.

Difficulty 1: we have little direct evidence about the genetics of behaviour in humans; and we know no way to obtain it for the specific behaviours that figure most prominently in sociobiological speculation—aggression and conformity, for instance. With our long generations, it is very difficult to amass much data on heritability. More importantly, we cannot (ethically, that is) perform the kind of breeding experiments,

How the elephant got its trunk

National Trust

in standardised environments, that would yield the required information. Thus, in dealing with humans, sociobiologists rely even more heavily than usual on speculative storytelling.

At this point, the political debate engendered by sociobiology comes appropriately to the fore. For these speculative stories about human behaviour have broad implications and proscriptions for social policy—and this is true quite apart from the intent or personal politics of the storyteller. Intent and usage are very different things; the latter marks political and social influence, the former is gossip or, at best, sociology.

The common political character and effect of these stories lies in the direction historically taken by nativistic arguments about human behaviour and capabilities—a defence of existing social arrangements as part of our biology.

In raising this point, I do not act to suppress truth for fear of its political consequences. Truth, as we understand it, must always be our primary criterion. We live, because we must, with all manner of unpleasant biological truths—death being the most pervasive and ineluctable. I complain because sociobiological stories are not truth, rather they are unsupported speculations with political clout (again, I must emphasise, quite apart from the intent of the storyteller). All science is embedded in cultural contexts, and the lower the ratio of data to social importance, the more science reflects the context.

In stating that there is politics in sociobiology, I do not criticise the scientists involved in it by claiming that an unconscious politics has intruded into a supposedly objective enterprise. For they are behaving like all good scientists—as human beings in a cultural context. I ask only for a more explicit recognition of the context—and, specifically, for more attention to the evident impact of speculative sociobiological stories. For example, when *The New York Times* runs a week-long front page series on women and their rising achievements and expectations, spends the first four days documenting their progress towards social equality, devotes the last day to potential limits upon this progress, and advances sociobiological stories as the only argument for potential limits—then we know that these are stories with consequences: "Sociologists believe that women will continue for some years to achieve greater parity with men, both in the work place and in the home. But an uneasy sense of frustration and pessimism is growing among some advocates of full female equality in the face of mounting conservative opposition. Moreover, even some staunch feminists are reluctantly reaching the conclusion that women's aspirations may ultimately be limited by inherent biological differences that will forever leave men the dominant sex" (*The New York Times,* 30 November, 1977).

The article then quotes two social scientists, each with a story. First: "If you define dominance as who occupies formal roles of responsibility, then there is no society where males are not dominant. When something is so universal, the probability is—as reluctant as I am to say it—that there is some quality of the organism that leads to this condition." Secondly: "It may mean that there never will be full parity in jobs, that women will always predominate in the caring tasks like teaching and social work and in the life sciences, while men will prevail in those requiring more aggression—business and politics, for example—and in the 'dead' sciences like physics."

Difficulty 2: the standard foundation of Darwinian just-so stories does not apply to humans. That foundation is the implication: if adaptive, then genetic—for the inference of adaptation is usually the only basis of a selective story, and Darwinism is a theory of genetic change and variation in populations.

Much of human behaviour is clearly adaptive, but the problem for sociobiology is that humans have developed an alternative, non-genetic system to support and transmit adaptive behaviour—cultural evolution. (An adaptive behaviour does not require genetic input and Darwinian

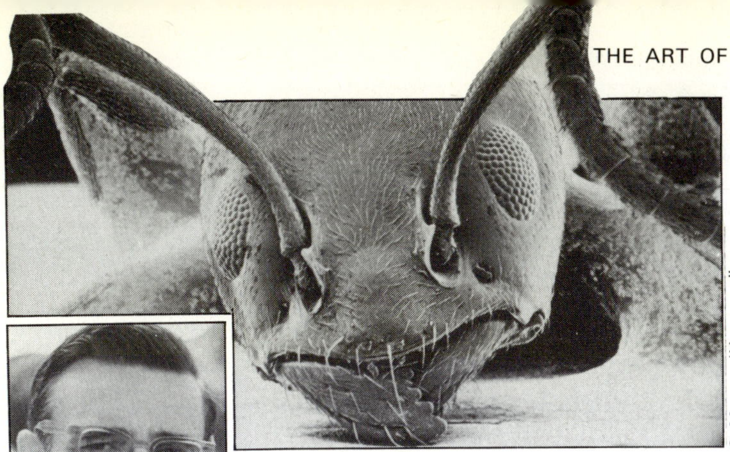

C. Morrow/Honeywell

E. O. Wilson, midwife to sociobiology, with one of his favourite species, a red ant

selection for its origin and maintenance in humans; it may arise by trial and error in a few individuals who do not differ genetically from their groupmates in any way relevant to this behaviour, spread by learning and imitation, and stabilise across generations by value, custom and tradition.) Moreover, cultural transmission is far more powerful in potential speed and spread than natural selection—for cultural evolution operates in the "Lamarckian" mode by inheritance through custom, writing and technology of characteristics acquired by human activity in each generation.

Thus, the existence of adaptive behaviour in humans says nothing about the probability of a genetic basis for it, or about the operation of natural selection. Take, for example, Trivers's concept of "reciprocal altruism". The phenomenon exists, to be sure, and it is clearly adaptive. In honest moments, we all acknowledge that many of our "altruistic" acts are performed in the hope and expectation of future reward. Can anyone imagine a stable society without bonds of reciprocal obligation? But structural necessities do not imply direct genetic coding. (All human behaviours are, of course, part of the potential range permitted by our genotype —but sociobiological speculations posit direct natural selection for specific behavioural traits.) As Benjamin Franklin said: "Either we hang together, or assuredly we will all hang separately".

The grandest goal—I do not say the only goal—of human sociobiology must fail in the face of these difficulties. That goal is no less than the reduction of the behavioural (indeed most of the social) sciences to Darwinian theory. Edward Wilson presents a vision of the human sciences shrinking in their independent domain, absorbed on one side by neurobiology and on the other by sociobiology.

But this vision cannot be fulfilled, for the reason cited above. Although we can identify adaptive behaviour in humans, we cannot tell if it is genetically based (while much of it must arise by fairly pure cultural evolution). Yet the reduction of the human sciences to Darwinism requires the genetic argument, for Darwinism is a theory about genetic change in populations. All else is analogy and metaphor.

My crystal ball shows the human sociobiologists retreating to a fallback position—indeed it is happening already. They will argue that this fallback is as powerful as their original position, though it actually represents the unravelling of their fondest hopes. They will argue: yes, indeed, we cannot tell whether an adaptive behaviour is genetically coded or not. But it does not matter. The same adaptive constraints apply whether the behaviour evolved by cultural or Darwinian routes, and biologists have identified and explicated the adaptive constraints.

But it does matter. It makes all the difference in the world whether human behaviours develop and stabilise by cultural evolution or by direct Darwinian selection for genes influencing specific adaptive actions. It makes a great difference because cultural and Darwinian evolution differ profoundly in the three major areas that embody what evolution, at least as a quantitative science, is all about:

1. Rate: cultural evolution, as a "Lamarckian" process, can proceed orders of magnitude more rapidly than Darwinian evolution. Natural selection continues its work within *Homo sapiens,* probably at characteristic rates for change in large, fairly stable populations, but the power of cultural evolution has dwarfed its influence (alteration in frequency of the sickling gene *vs* changes in modes of communication and transportation). Consider what we have done in the past 3000 years, all without the slightest evidence for any change in the power of the human brain.
2. Modifiability: complex traits of cultural evolution can be altered rapidly; Darwinian change is limited to much slower rates of spread of alleles by natural selection.
3. Diffusability: since traits of cultural evolution can be transmitted by imitation and inculcation, evolutionary patterns include frequent and complex anastomoses among branches. Darwinian evolution is a process of continuous divergence and ramification.

I believe that the future will bring mutual illumination between two vigorous, independent disciplines—Darwinian theory and cultural history. This is a good thing, joyously to be welcomed. But there will be no reduction of the human sciences to Darwinian theory and the research programme of human sociobiology will fail. The name, of course, may survive. It is an irony of history that movements are judged successful if their label sticks, though the emerging content of a discipline may lie closer to what opponents originally advocated. Modern geology, for example, is an even blend of Lyell's strict uniformitarianism and the claims of catastrophists. But we call the hybrid doctrine by Lyell's name.

I welcome the coming failure of reductionistic hopes because it will lead us to recognise human complexity at its proper level.

For consumption by *Time's* millions, my colleague Bob Trivers maintained: "Sooner or later, political science, law, economics, psychology, psychiatry and anthropology will all be branches of sociobiology" (*Time,* 1 August, 1977). It is one thing to conjecture, as I would allow, that common features among independently developed legal systems might reflect adaptive constraints and might be explicated usefully with some biological analogies. It is quite another to state, as Trivers did, that the legal profession, among others, will be subsumed as mere epiphenomena of Darwinian processes.

I read Trivers's statement the day after I had sung in a full production of Berlioz's *Requiem.* And I remembered the visceral reaction I had experienced upon hearing the four brass choirs, finally amalgamated with the 10 tympani in the massive din preceding the great *Tuba mirum*—the spine tingling and the involuntary tears that almost prevented me from singing. I tried to analyse it in the terms of Wilson's conjecture—reduction of behaviour to neurobiology on one hand and sociobiology on the other. And I realised that this conjecture might apply to my experience. My reaction had been physiological and, as a good mechanist, I do not doubt that its neurological foundation can be ascertained. I will also not be surprised to learn that the reaction has something to do with adaptation (emotional overwhelming to cement group coherence in the face of danger, to tell a story). But I also realised that these explanations, however "true", could never capture the meaning of that experience.

And I say this not to espouse mysticism or incomprehensibility, but merely to assert that the world of human behaviour is too complex and multifarious to be unlocked by any simple key. I say this to maintain that this richness—if anything—is both our hope and our essence.

The full version of this article was published in *Sociobiology: Beyond Nature/Nurture,* edited by George W. Barlow and James Silverberg, AAAS Selected Symposium No. 35, Westview Press, Boulder, Colorado.

PART VI
LAMARCK IS NO ALTERNATIVE

Lamarck came up with a theory to explain evolution half a century before Darwin. Lamarckism is, however, unworkable. Nevertheless, from time to time evidence of apparently Lamarckian inheritance surfaces. In this section we reprint our examination of the most recent outbreak and follow attempts to reproduce the extravagant claims. They failed, and there are good reasons to suppose that Lamarckism must fail as an explanation of adaptation

Chapter Seventeen

Lamarck lives—in the immune system

An Australian immunologist produced results which suggested that animals could inherit characteristics acquired by their parents. This early appraisal examines the work and its implications

COLIN TUDGE

IT IS a truth universally acknowledged that the mechanism of evolution described by Jean Baptiste de Lamarck in the 18th century was wrong. He proposed that the characteristics that animals acquire during their life in response to life's pressures—the neck of the giraffe, for instance, elongated by stretching for the top-most leaves—were passed on to the offspring. In his own lifetime Lamarck's ideas excited ridicule, and Charles Darwin's quite different idea put forward a half-century later—that evolutionary change occurs by selection of favoured individuals among a variety of offspring—has become, with refinements, the rubric of 20th-century biology.

Odd then, that an Australian immunologist has recently been presenting a phenomenon that apparently can be explained only in Lamarckian terms; indeed his research was specifically designed to test the idea that some characteristics that can be acquired only in an animal's life, and which could be passed on to the offspring only through the genes (and not, say, by infection) are indeed inherited. The immunologist, Ted Steele, is taken seriously; and if he is right, we are beginning a new chapter in biology.

Steele explains the background to his ideas in *Somatic Selection and Adaptive Evolution,* which he wrote in the winter of 1978, when he worked at the Ontario Cancer Institute. His argument focuses upon, and seems to overthrow, the ideas of August Weismann, one of the first German scientists to declare support for Darwin in the late 19th century. Weismann declared that the genes within the germ cells—the ova and sperms, which are the only cells that pass on information from generation to generation—are quite independent of the genes within the somatic cells (the cells, that is, of the rest of the body). Weismann's idea apparently provides a simple physical reason for rejecting Lamarckism. Body cells will be modified in response to the environment, obviously; a blacksmith develops big muscles, and calloused hands. And these physical changes may well be reflected in the activity of the genes within his muscle cells; after all, a blacksmith cannot develop big muscles unless his genes issue directions to manufacture more protein. But the somatic cells are separate from the reproductive cells. The blacksmith's muscles bloom like flowers in the desert; there is (says Weismann) no mechanism that could feed back information on the changes within them, to the genes within his spermatozoa. However much the blacksmith toils his children must work themselves if they too want muscles. And if smithing really does in the end produce a muscular stock, it would be for Darwinian reasons; that the offspring who really had no capacity to produce big muscles were selected out, and drifted into other trades.

Keystone

A blacksmith, big and muscular, but his children might be weaklings

In obvious and easily quantifiable cases like this, Weismann's doctrine holds out. It is always arrogant in biology to suggest that any one thing *cannot* happen, and astonishingly bold of Weismann to suggest at the end of the 19th century, when cell physiology was primitive and molecular biology had not begun, that there could be no feedback between somatic and germ cells. But in cases like that of the muscular blacksmith and in a million other similarly banal examples one might dream up, his doctrine seems to hold; and as Steele says, it plays "an axiomatic role in modern evolutionary theory".

Weismann's line of thought takes a subtler twist with the advent of Francis Crick and James Watson, from the 1950s onwards. They, as everyone knows, worked out the structure of DNA, the stuff of which genes are made. They also contributed much to what has become the *leitmotiv* of molecular biology: that DNA, coding the basic instructions of the cell, makes RNA, which in turn makes protein. And to this Crick and Watson have appended their "central dogma": that the flow of information is in one direction only—or, more specifically, that there is no feedback of information from the *protein* to the DNA. It is the protein (and all the other bodily components to which the special proteins known as enzymes give rise) that is the stuff of flesh; and it is the flesh that rubs against the environment and is modified by it. If the flesh can be shown at the molecular level not to feed back its "knowledge" to the central store of knowledge in the DNA, then again Lamarckian ideas begin simply to seem unbiological. The DNA turns out proteins and these in turn are knocked around by life; but they do not return to the DNA and share the light of their experience. The DNA operates as the neodarwinians expect: a blind general far behind the lines, cut off from the troops, sending out more forces to carry out tasks of which he has no direct intelligence.

Crick and Watson's central dogma has been supported experimentally many times. It is germane to Steele's argument because at first sight (and to many biologists) the Crick and Watson dogma looks like a restatement, at the molecular level, of Weismann's doctrine. In both cases there is, it seems, no feedback: none from the cell itself which worked out the instructions of the DNA, to the DNA that gave those instructions; and none (as Weismann said) from the somatic cells which effectively enact the instructions encoded in the sperms and ova—instructions that will be passed on to the next generation.

But as Steele points out, there is no direct connection between Weismann's doctrine, which is old and bold, and hypothetical, and Crick and Watson's dogma which is modern and well worked out, and deserves to be believed. We know (which the Crick and Watson dogma does not deny) that RNA feeds back information to DNA, and that the DNA can incorporate this information. Why should not RNA that is produced within the somatic cells be fed back into the DNA within the sex cells, to become part of the instruction that is inherited? This is a kind of Lamarckism: the experience of the body (made manifest in RNA) fed directly into the heritable genes. And it is not unbiological. The mechanisms for feeding new genetic material into DNA are well established; they are the stuff of the modern craft of genetic engineering. Viruses, which carry nucleic acid in a protected state by coating them in protein (protected, that is, against the body enzymes that attack free nucleic acid), are one of several possible vehicles that could carry somatic RNA to the DNA of the sex cells.

The implications of Steele's idea are stupendous. But the point, before we discuss those implications, is that his idea is testable. All that is needed is to find a system within the body that is known (or may be supposed) to produce quite new genetic material in the course of an animal's life; and then to show that this novel material is indeed fed back into the sex cells, and is passed on to succeeding generations. There is (or may be supposed to be) such a system. It comprises the lymphocytes and stem cells within the bone marrow that produce antibodies: the cells of the immune system.

The lymphocytes appear in the blood as white blood cells, but they are produced from stem cells in the bone marrow. Their task is to identify and repel material foreign to the body—bacteria, viruses or (as transplant surgeons are only too aware) transplanted organs. They perform this task in many ways, and one way is to produce antibodies. Antibodies are proteins, and they operate, initially, by attaching themselves like locks around keys on to foreign-looking entities on invading materials, known as antigens. But the apparent simplicity of the mechanism conceals deep conceptual problems.

The first is how it is that the body does not, at any one time, produce or have in store all the antibodies that it might need. The first time a baby is infected with measles virus, for example, it seems to have no appropriate antibodies. Only with time does it develop them, and then repels the infection. But in subsequent infections the baby will demonstrate its immunological memory and greets any fresh invasion with the full might of a specifically primed army of antibodies.

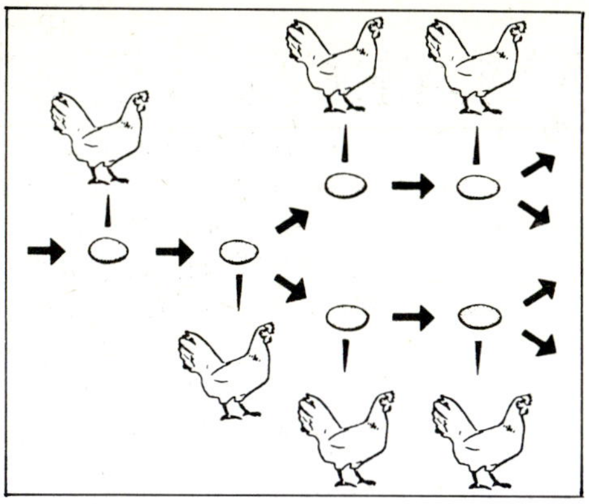

Samuel Butler said that the chicken is the egg's way of making another egg. This neatly illustrates the distinction between germ-line (egg) and soma (chicken)

This ability does not seem so difficult to explain. We need to suppose only that different kinds of lymphocytes produce different kinds of antibody. To produce all possible antibodies all the time would be ludicrously wasteful. Instead, the body waits for a particular antigen to appear and only then pushes out appropriate antibodies, like a factory satisfying a sudden demand. And the system remains tooled up; with subsequent demand it can swing into service at the first sign of trouble.

But the second problem is the sheer range of antibodies that the immune system can produce. Some foreign materials—some plastics used in surgery for example—do not evoke an immune response. But most solid and complex components of the world, put into an animal, will provoke some kind of antibody response: and usually several kinds of antibody will be invoked by one foreign substance. Indeed, immunologists calculate that any one man or mouse can produce 10^7 or so different kinds of antibody; that is, 10 million different kinds of protein with the single task of repelling invaders.

Biologists have sought to explain this astonishing versatility in two main ways. The first, the so-called template theory, is intuitively the more attractive. It supposes that an antibody is like wax; that it is produced in a fairly crude form and then moulded around the antigen; and that what constitutes the immunological memory is the memory of the mould, like a die in a machine-tool workshop. The idea is attractive because it is economical, and is, after all, the method that any sensible factory owner would adopt, if faced with a comparable problem.

But intuitive or not, the template idea does not work. For one thing, antibodies are not like wax. Antibodies are proteins, and proteins consist of chains of amino acids. The effectiveness of the antibody depends upon how well it fits onto the corresponding antigen, which depends upon the shape of the antibody. But an antibody will take one form and one form only, depending on the sequence of amino acids within its chains. In short, all the different 10^7 possible antibodies are different chemically. They must all, therefore, it seems, be separately coded by different kinds of RNA.

Thus, the attractive template theory has had to give way to the superficially much less attractive idea of clonal selection. This says that for each of the different potential kinds of antibody there must be a different kind of lymphocyte—each capable of producing one kind of antibody, and one only. What a particular kind of antigen does is to stimulate the appropriate kind of lymphocyte to reproduce; and all the lymphocyte's offspring, all identical and collectively called a clone, produce the antibodies which, given just a little time, can mount an effective attack on the particular antigen. The antigen thus selects the clone that brings about its own destruction.

But this idea of "clonal selection" is intuitively difficult. If an animal can produce 10^7 different kinds of lymphocyte then it must inherit a vast library of 10^7 appropriate genes from its parents. Furthermore, a given generation of animals will be able to produce antibodies against antigens to which its own ancestors have never been exposed. The necessary library of genes is not only large, but is remarkably conservative; the genes remain in stock even when they are almost never called upon. Steele suggests in *Somatic Selection and Adaptive Evolution* that such conservatism, in the absence of selection, is not usually observed among other genes coding for other kinds of characteristics. So Steele invokes a third hypothesis. It appears Lamarckian, but it is not; it is straightforward Darwinism, applied to cells. The Lamarckian twist comes later.

This third hypothesis supposes that although the lymphocytes —or rather the stem cells that give rise to lymphocytes—have enormous versatility, they do not at any one time contain the complete range of 10^7 varieties that we know can be invoked. Instead, the population of lymphocytes behaves, within the environment of the body, just as Darwin supposed a population of animals behaves within the environment of its habitat. It throws up, all the time, a great variety of mutant types, each capable of producing just one antibody. Most of these types meet no corresponding antigen, and they die; as if an animal adapted to a particular esoteric diet found itself without appropriate food. But if the particular mutant meets its corresponding antigen, then it is stimulated to reproduce. Then it produces a clone. Then, and only then, does the particular mutant become established permanently in the immunological repertoire of the particular individual.

This hypothesis supposes nothing less than small-scale Darwinian evolution within the immune system. Mutation produces variety; natural selection, exercised by extraneous antigens, determines which of the varieties survives. But the point is that variety here means genetic variety. Each surviving lymphocyte, when once sanctified by the stimulated act of cloning, is a new genotype.

According to the third hypothesis of immunological versatility, when an animal mounts an antibody response against a novel antigen, it can do so only by selecting out a novel clone of lymphocytes; it must be selecting out a kind of RNA that none of its ancestors necessarily saw fit to establish. And suppose, as Steele also suggests, that this new kind of RNA could somehow be transported into the germ cells of that animal. Then some of that animal's descendants—not all, but a proportion—should be able to produce the same response to the antigen as the parent.

This is what Steele and his colleague Reg Gorczynski seem to have demonstrated. They reported their first results in 1980, in the *Proceedings of the National Academy of Sciences* (vol 77, p 2871). Their work was on the phenomenon of tolerance—a state in which an animal's immune system is treated so that it *fails* to produce antibodies against particular foreign antigens. The mechanism of tolerance seems to depend upon selection of specific clones of suppressor cells (lymphocytes processed by the thymus gland—that is, T-cells —which counteract the activity of reactive clones. Tolerance is not brought about merely by *absence* of reactive lymphocytes). Gorczynski and Steele, then at the Ontario Cancer Institute in Toronto, rendered CBA strain mice tolerant to antigens from A/J mice by injecting them repeatedly with lymphoid cells from CBA-A/J crossbred mice. Then they showed that more than half of the offspring of the tolerant CBA mice—offspring of tolerant fathers, that is—were also tolerant.

Nine months later, in 1981, Gorczynski (still at Toronto) and Steele (then at the Medical Research Council's Clinical Research Centre) reported a significant extension of that first finding (*Nature*, vol 289, p 678). They induced a state of tolerance in B10 mice, to *two* different antigens from two different mouse strains, and then showed that this tolerance could be passed from tolerant fathers to a proportion of offspring. Furthermore the offspring sometimes inherited their father's tolerance to only one antigen, and sometimes to both—suggesting that there was some kind of random filtering of the acquired characteristic into the germ line.

Thus we seem to have a Lamarckian system. Through being exposed to a novel antigen, mice produced an appropriate clone of antibodies containing a novel RNA. The information within this RNA—from somatic, lymphocyte cells—then (the hypothesis suggests) became incorporated within the DNA of their germ cells. And then it was passed on to the offspring.

But perhaps this all seems too particular. The immune system, after all, seems special. No other tissue in the body, surely, operates with such extraordinary profligacy—throwing out thousands of mutants to be culled by the internal environment.

But the immune system is possibly not so different as we have been brought up to believe. It is at least possible that the lymphocyte's dynamic approach to life, its constant speculative probing of its surroundings with new mutant cell types, most of which will peter out but some of which will be favoured, is the *modus operandi* (albeit in quieter form) of all tissues. Steele, in his book, mentions epithelial cells of the body's surfaces, like those of the skin and gut. Such cells (like the lymphocytes) have an extraordinarily rapid turnover. They also (like the lymphocytes) are exposed to an astonishing variety of environmental influences: in the case of the lymphocyte, an antigen; in the case of a gut cell, some recondite component of diet, such as plant toxin. We know that populations of animals (and humans) adapt to odd diets which others cannot tolerate. An economical way of explaining this adaptation would be to suppose that the epithelial cells constantly produce mutant types—some of which survive the assaults of the potential toxin, some of which do not. Those that do survive, in simple Darwinian fashion, give rise to clones, which eventually dominate the tissue. If we then suppose (as Steele seems to have demonstrated in the immune system) that some of the *novel* genetic information in these favoured mutants—which become numerically prevalent —filtered back into the animal's germ cells, then we have an acquired, heritable characteristic. Then we have Lamarckism.

So let us return to the elongation of the neck of the giraffe which (for reasons that are largely spurious) is what Lamarck tends to be remembered for. Some anti-Darwinians have argued that the giraffe's neck is such an extravagant device, and must be accompanied by so many other adaptations—including a system for adjusting blood flow to the head to avoid fainting when the animal drinks—that it could not have evolved simply by natural selection acting upon chance mutations; that there must have been some guiding hand. Modern Darwinians have a way of knocking such arguments for six; mutation and natural selection, combined with ideas on gene recombination during sexual reproduction, and the general flow of genetic information within a population, can explain the giraffe adequately. There is no need for *ad hoc* hypotheses.

But to say that there is no need to propose non-Darwinian mechanisms does not mean that nature has not produced such mechanisms. Steele's idea, applied to the neck of the giraffe, would be no more than a refinement—albeit an important one—of Darwinian selection. It would suppose that as a giraffe stretched its neck, so it would, within its own body, select out clones of cells compatible with stretching; and that such clones would eventually dominate. Then, as seems to be happening within the immune system of his and Gorczynski's tolerant mice, some of the RNA information from those favoured clones would filter back into the animal's germ cells. Thus in addition to the natural selection proposed by Darwin, Steele proposes a neo-Lamarckian tracking system. If such a system exists then evolution ceases to be quite the blind agency that is supposed.

All of the above is an attempt simply to give an outline of Steele's ideas. The description is not advocacy; these few results and accompanying maverick hypothesis are not enough to initiate a counter-reformation. First, biology does not need a neo-Lamarckian system. The Darwinian enigmas

No amount of stretching will produce long-necked offspring

are apparent rather than real. But if such a system does exist, then it cannot be ignored.

Secondly, to extrapolate from the immune system to other systems of the body is an athletic leap. There is no unequivocal experimental evidence that phenomena such as Steele describes are found in other tissues. But to say that no evidence exists for other systems is merely to say that the experiments have not been done; and they have not been done for two sound reasons. The first is that few career biologists would care to attempt a serious Lamarckian experiment (it just is not done); and the second is that in order to demonstrate such phenomena the biologist must choose a system that is extremely well understood at the genetic level, and which produces readily observable and repeatable phenomena. These conditions do apply in the immune system, but do not apply, so far, to most other systems. But it is theoretically possible to test Steele's ideas in other systems.

Then again, although the immune system does appear to behave differently from other systems, there is no good reason to suppose it really is different; indeed, it *feels* more biological to suggest that the immune system's method of cloning and self-selecting is an exaggeration of a general method of operating, than to suggest that it has invented a completely new *modus operandi.* Steele himself puts the issue the other way round. Immunologists, he says, are trained to think that the system they deal with "remembers", and adapts to circumstance. Other kinds of biologists are not so trained, and for them to think in such terms would require a "paradigm shift". But if they applied his kind of ideas to data they already have, might they not see comparable phenomena? And how many experiments—in genetics especially—are conveniently forgotten because they produce unexpected results? Would it not be reasonable at least to be alert to the possibility that mechanisms comparable with those proposed by Ted Steele may sometimes be at work?

Finally, and perhaps most importantly, Steele, beyond any question, is an excellent practical scientist, which is why he is invited to work in some of the world's most prestigious laboratories. He is currently repeating the results he produced in Toronto, at the Clinical Research Centre in London*. Whether or not the phenomena he describes will in the end make a serious impact on evolutionary theory, they should be taken seriously and should be explored, both within the immune system (with immune responses other than tolerance) and within other tissue systems.

*This was in March 1981; for the outcome of that work see the next chapter.

Chapter Eighteen

Lamarck revisited—the debate goes on

Other scientists could not repeat Ted Steele's experiments, and called his Lamarckist claims into question. The conclusion seems to be that Steele was mistaken

MIRANDA ROBERTSON

SIX weeks after Gorczynski's and Steele's paper appeared in *Nature* in February 1981, the journal published two different papers from highly reputable biologists reporting two separate failures to demonstrate the same phenomenon (vol 290, pp 508 and 513). Two of the biologists involved are Sir Peter Medawar and Leslie Brent, who first developed the means of inducing tolerance used by Steele and Gorczynski—and who in fact looked for Lamarckian inheritance in their own early experiments and failed to find it. They have now failed to find it again, and they and their research collaborators have spent the best part of a year in the fruitless attempt. It is the fear of precisely such a disappointing outcome that often deters scientists from following up their colleagues' more revolutionary reports. In this chapter I shall first suggest the sort of reason why most of the laboratories equipped to follow up Gorczynski's and Steele's experiments did not bother to do so; then I shall try to explain how Medawar and his collaborators, working on the same strains of mice and using almost the same techniques, could have arrived at such very different conclusions. Competent scientists carefully weigh the importance of the result against the chances of its being right. Because Steele's experiment was devised to test a particular theory about the mechanism of adaptive change, the plausibility of the theory is an important factor in the balance.

Everybody, including Steele, accepts that organisms inherit their characteristics through the DNA contained in their parents' germ cells—the eggs and sperm. No matter how favourable any adaptive change that may occur in a parent during his or her lifetime, it will not be passed on to the offspring unless it induces a corresponding change in the DNA of a germ cell. Assuming the adaptive change entails a mutation in the DNA, there is a route by which, at least in theory, it could be transmitted to the germ cells. This route involves a kind of virus, known as a retrovirus, which can pick up fragments of DNA from one cell and insert them into another; and it was first suggested by Howard Temin, who won a Nobel prize for his work on retroviruses. (Incidentally it was Temin who submitted Steele's first Lamarckian paper to the *Proceedings of the National Academy of Sciences*.)

The retrovirus is not likely to be a particularly reliable vehicle, and the chances of a particular mutant gene's finding its way into an egg or a sperm will probably be proportional to the number of cells carrying the gene. That is what makes the immune system a particularly good one in which to look for Lamarckian inheritance: for the production of antibodies against an infection involves mutation followed by massive proliferation of the mutant cells that produce the antibodies. Curiously, Steele's own experiments were not concerned with antibodies. He was looking at a different and much less well understood aspect of the immune response; but we shall come to that later. For the time being, let us examine the sort of general objections people have raised to the theory.

There are two quite simple ones. First, if mutant antibody genes can find their way into the germ-line, even occasionally, then one would expect resistance to endemic diseases to spread rapidly through populations in the less developed parts of the world. Secondly, the pattern of inheritance of genetic traits should be much less predictable than it is. On the whole, our genetics obey the laws Gregor Mendel formulated on the basis of his celebrated experiments on peas. If mutant genes were being shipped into our germ-lines at any appreciable rate by retroviruses we should expect to see obvious departures from those rules.

Other, more technical objections have been raised to more specific details of Steele's arguments, but it would take too long to explain them here. Besides, objections can be raised to aspects of any theory without necessarily invalidating it (the theory of natural selection is a case in point). What has discouraged many critics from taking Steele seriously is the inadequacy of his own critical examination of his ideas—as they see it reflected in his book.

It has not, however, discouraged everyone. In particular, the book was favourably reviewed by two distinguished academics: the geneticist James Crow and Sir Peter Medawar. By the time the book was published, Steele, in collaboration with Gorczynski at the University of Toronto, had produced evidence in support of his theory; and Medawar was sufficiently impressed and interested to invite Steele to work in his laboratory at the Clinical Research Centre in Harrow.

Steele's experiments were not concerned with antibody production, which is an adaptive response that is known to entail the proliferation of mutant cells. Instead, he looked at an experimental artefact known as acquired neonatal tolerance whose adaptive significance is not entirely clear and whose mechanism is uncertain.

The first of the papers published in *Nature* reported an attempt by Leslie Brent and Lee Rayfield at St Mary's Medical School, London, and Philip Chandler, Walter Fierz, Sir Peter Medawar and Elizabeth Simpson at the Clinical Research Centre in Harrow, to replicate the original experiment of Gorczynski and Steele as closely as possible. (Although one of Steele's crucial arguments is that Brent *et al* did not replicate his protocol precisely enough.) The results of this experiment show no evidence at all of the inheritance of tolerance. Assuming that neither group is either actually crooked or hopelessly incompetent, how could such different results have come out of such similar experiments?

The answer may lie in the nature of the tests for tolerance. Gorczynski and Steele did not actually use skin grafts, which would have been the true acid test, because skin evokes a particularly vigorous rejection response in a non-tolerant animal. Instead, they used a laboratory measure of immune reactivity that is believed to be, so to speak, a test-tube counterpart of graft rejection.

The cells that engage in the various battles undertaken by the immune system are the white blood cells, or lymphocytes. The first step in the test, therefore, is to remove the lymphocytes from samples of blood or spleen. There are several different kinds of lymphocyte; one is the kind that produces

A negative end for acquired tolerance

AT THE time Miranda Robertson was writing, Steele's theory was suffering from both a failure to repeat his own results, and a refutation in an independent test. The independent test by Anne McLaren and her colleagues he did not comment on. The repetition of his experiment by Leslie Brent and his colleagues, however, has a sequel.

The Brent team's results looked undeniably negative: no inheritance of acquired tolerance. But Steele himself was to take the unlikely course of denying their conclusions. He argued, in the *New Scientist* (vol 90, p 360), that "contrary to their announcements, Brent and his colleagues have in fact, and in their own way, reported positive results supporting the initial findings of Gorczynski and myself". Steele had been re-analysing Brent *et al*'s original data, and had persuaded himself that, despite all appearance to the contrary, their results substantiated his theory.

Steele did not persuade Brent and his colleagues, who soon destroyed the re-analysis (*New Scientist*, vol 90, p 493). The exact details of the exchange need not detain us: Steele had selected only those parts of the results that fitted his theory best, and used only those selected parts in a statistical test. The test supported his theory only because he had excluded results that disagreed with it. When all the results were considered, the original conclusion still stood: no inheritance of acquired tolerance. Many immunologists, however, felt a nagging discomfort about a minor difference of technique between the original experiment and its repeat. The repetition was not *exact*. The difference lay in the dose and kind of cells used to make the parental mice tolerant. Gorczynski and Steele had used a dose of 10^8 cells; Brent *et al* but 5×10^7. And, if you please, the cells used by Gorczynski and Steele were spleen cells and bone marrow cells in a ratio of 9:1; Brent *et al* had used 1:1. The difference is tiny, and Brent *et al* argued that anyway it would not cause the difference in the results. But the really definitive solution was to repeat the original Gorczynski and Steele experiments as accurately as possible. That would mean doing the whole experiment again, using the same sorts of cells, in the same quantities, that Steele had used.

The Brent team has since done an exact repeat. The results were again negative (*Nature*, vol 295, p 242). All the independent repeats and trials of Gorczynski's and Steele's ideas have therefore been negative. For the time being at least, we can conclude with Brent and his colleagues that acquired immunological tolerance is not inherited.

Mark Ridley

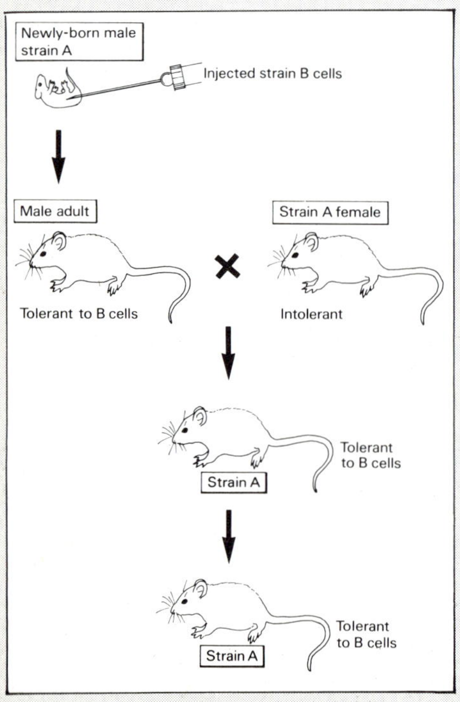

How mice are believed to pass on tolerance to their offspring

But on balance it seems that mice cannot pass acquired tolerance to their offspring

antibodies—but antibodies are not believed to play a very important part in graft rejection. The main protagonists in rejection are known as T lymphocytes, some of which are specialised to attack and kill the cells of the graft. These attackers can be aroused in culture dishes by exposure to cells of a different strain. So the next step in the tests used both by Gorczynski and Steele, and by the collaborating team at St Mary's and the Clinical Research Centre, is to expose lymphocytes taken from the offspring of the tolerant fathers to inactivated lymphocytes collected from the mice used to induce tolerance in their fathers. This sensitises the cells unless the animal itself is tolerant, and its cells do not become sensitised.

Sir Peter Medawar, at a symposium held to honour him in July 1980

To find out whether the offspring cells have been sensitised, they are introduced a second time to target cells, which this time have been marked with a radioactive label. If they are attacked and killed by the offspring cells, they will disintegrate and release their label into the culture fluid. So by measuring radioactivity in the culture medium, the immunologist can measure the magnitude of the attack on the target cells.

Several factors will influence the outcome of such tests. One, for example, is the ratio of attacking to target cells. Another is the particular strain of mouse used, or even the individual mice. Other factors may be harder to identify. For example, for reasons that are not clear, the radioactive release even in Steele's control experiments was generally rather low. All in all, such tests are subject to more or less random fluctuation, and that means that controls and the statistical testing of the results must be rigorous. It also means—a point emphasised by Medawar—that different experimenters working with the same system might occasionally arrive at different results just by chance.

Elizabeth Simpson at the Clinical Research Centre is a particular expert in these tests, and Brent and his co-workers were careful to eliminate chance effects as far as possible. They used large numbers of mice and made comparisons between mice in different litters, as well as comparing those in the experimental groups with similar mice with non-tolerant fathers, to make sure any unresponsiveness was really due to tolerance and not just to natural variation within the strain. They also conscientiously avoided bias by reading the radioactivity measurements "blind"—that is, without knowing until afterwards which mouse was being tested.

They took one further important step: they carried out skin-graft tests, both on the fathers and on their progeny. The fathers accepted their grafts permanently, the progeny rejected theirs within days.

Then there was a second and rather different kind of experiment, also reported in *Nature*, demonstrating a total failure to show Lamarckian inheritance. For that experiment, Simpson's CRC team worked with Anne McLaren, an embryologist at University College, London, who has perfected techniques for making chimeric mice. A chimera, in this technical sense, is a mouse made by removing two very early mouse embryos from mothers of different strains, fusing them together, and returning them to the uterus of a third mouse to complete their development. Such fused embryos grow up into normal mice with a mixture of tissues derived from the cells of the two original embryos. A consequence of the extreme intimacy of the two genetically different strains of cell at a formative stage of development is that they are immunologically tolerant of one another. Which is just as well for the mouse.

But is this tolerance heritable? Apparently not. Chimeras mated with pure-bred animals of the parental strains will produce progeny of one strain or the other. The progeny will not tolerate cells of the opposite strain.

Steele, nonetheless, persists with his own experiments and insists that they continue to work in his hands: skin grafts are under way in Toronto but results are not yet available. Time will tell if he is right or wrong; but if he is right he may need a less complex system in order to demonstrate it.

In either case, it would be wrong to regard Steele as challenging Darwinian natural selection, although such a challenge seems to have been inferred by the media. First of all, as explained in the previous chapter, Steele would not deny the importance of natural selection.

Secondly, Darwin himself warmly espoused Lamarckian notions of the inheritance of acquired characteristics and indeed worked out in some considerable detail a theory not unlike Steele's own, which he called pangenesis. But while the validity of Darwin's theory of natural selection is not in question, it is now clear that his theory of pangenesis—in which he recognised serious weaknesses—is in fact wrong.

Chapter Nineteen

The necessity of Darwinism

There is no evidence for Lamarckian inheritance, but even without evidence we can be reasonably sure that the inheritance of acquired characteristics just will not do

RICHARD DAWKINS

BIOLOGY is the study of the complex things in the Universe. Physics is the study of the simple ones. It is the complexity of life, coupled with the precision of its adaptation, that cries out for a special kind of explanation, and the hunger for such explanation has frequently driven people to believe in a supernatural Creator. Complexity means statistical improbability. The more statistically improbable a thing is, the less can we believe that it just happened by blind chance. Superficially the obvious alternative to chance is an intelligent Designer. But Charles Darwin showed how it is possible for blind physical forces to mimic the effects of conscious design, and, by operating as a cumulative filter of chance variations, to lead eventually to organised and adaptive complexity, to mosquitoes and mammoths, to humans and therefore, indirectly, to books and computers.

Darwin's theory is now supported by all the available relevant evidence, and its truth is not doubted by any serious modern biologist. But, important as evidence is, in this article I want to explore the possibility of developing a different kind of argument. I suspect that it may be possible to show that, regardless of evidence, Darwinian natural selection is the only force we know that *could*, in principle, do the job of explaining the existence of organised and adaptive complexity. At this stage the argument will not be rigorous in a formal logical sense. I will examine the only available alternatives to Darwinian natural selection, and show that, even if the evidence did not go against them, they could not do the conceptual job we ask of them. (A similar argument could be developed in favour of the necessity of gradualistic rather than saltationist or "jumpy" evolution, at least if saltation is interpreted in the way Darwin interpreted it.)

Ingram Pinn

I know of only two alternatives to Darwinism that are offered as explanations of the organised and apparently purposeful complexity of life. These are God and Lamarckism. I am afraid I shall give God short shrift. He may have many virtues: no doubt He is invaluable as a pricker of the conscience and a comfort to the dying and the bereaved, but as an explanation of organised complexity He simply will not do. It is organised complexity we are trying to explain, so it is footling to invoke in explanation a being sufficiently organised and complex to create it.

Lamarckism is more interesting, and I will devote some time to it; I hope I will not seem to be flogging a dead horse. It is true that there is not a shred of good evidence in favour of the Lamarckian hypothesis, despite numerous efforts by far from detached zealots. But that could be just a fact about life on this planet. Maybe in some other galaxy there is a form of life whose evolution has been guided by Lamarckian forces. In this chapter I shall be exploring the possibility that we do not even need to go and look. It may be that we can rule out Lamarckism on logical grounds.

Historians of Lamarck sometimes complain that modern writers unfairly over-emphasise parts of his theory—the principles of use and disuse, and of the inheritance of acquired characteristics—at the expense of other ideas which Lamarck himself thought more important, notably the idea that if an animal *needs* some organ that need will somehow call the organ into existence. But this notion is so obviously mystical to the modern mind that it is fairer to Lamarck for us to concentrate on those parts of his theory that at least seem to have *some* chance of explaining evolution. This means concentrating on use and disuse, and on the inheritance of acquired characteristics.

Acquired characteristics are of many kinds. Most are injuries, which would not help us to explain evolution even if they were inherited, since they are deteriorations not improvements. Other acquired characteristics can be regarded as improvements. Muscles that have been used grow big, and this makes them better in the future at doing the very job for which they have in the past been used. Skin which is exposed to wear, for examples the soles of the feet when we walk barefoot, tends to become tougher. Therefore, a person who walks barefoot becomes, as a result, better equipped to walk barefoot. If we expose ourselves to the Sun we go brown, which better equips us to withstand the harmful effects of the Sun's rays. Above all the nervous system, as a result of experience in the world, changes its state in such a way that it survives better in that world. An animal that has been stung by a scorpion acquires a permanent change in its nervous system that tends to make it avoid scorpions in the future.

So, acquired characteristics can be improvements. An animal is better equipped to survive in the future, because it has acquired some characteristics in the past. If only, the theory runs, an acquired improvement could be inherited by the animal's children, improvements could accumulate over the generations, and that would constitute progressive and adaptive evolution.

My first theoretical difficulty is over the question of how the system "decides" whether an acquired characteristic is an improvement. It may have been in response to this difficulty that Lamarck invoked his mystical "need" principle. As I have already remarked, most acquired characteristics are injuries, not improvements, and there has to be some mechanism by which the improvements are chosen for inheritance, rather than the injuries. It is all very well to assert that if you walk on your bare feet your soles get tougher, but why do they get tougher instead of weaker? The Darwinian has a ready answer: they get tougher because in the past those animals whose feet happened to have a genetic tendency to get weaker with wear did not leave many descendants. Selection can see to it that the characteristics that animals acquire during their lifetimes are improvements. Given this under-pinning of natural selection, it might well be nice if children could directly inherit the acquired characteristics of their parents, for those acquired characteristics could then, indeed, tend to be improvements. But the improvement comes from natural selection in the first place. All that Lamarckian inheritance, if it existed, could achieve would be to prolong the benefits over more than one generation.

The argument applies with

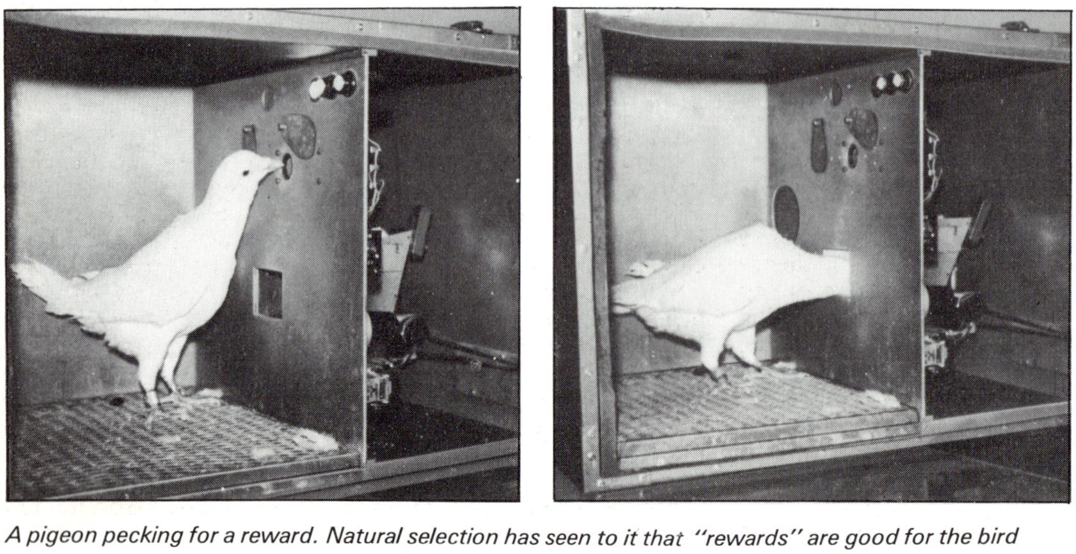
A pigeon pecking for a reward. Natural selection has seen to it that "rewards" are good for the bird

even more force to the case of behaviour, where there is a particularly important kind of acquired improvement—learning. An animal having learned something is normally in a better state than it was before it learned. A chick hatches out of the egg with a tendency to peck at almost anything small. After a while, if exposed to wheat grains mixed with sawdust, it learns to restrict its pecks specifically to grains rather than specks of sawdust. It has learned that wheat is good to eat and sawdust is not. In general, a learned behaviour pattern helps an animal to survive, and it is an acquired characteristic that might very profitably be handed on from parent to offspring, if only it were possible. (It *is* possible, of course, but not by genetic means.)

But again, what determines that the changes that animals undergo when they learn are changes in the direction of improvement? Why do animals learn to do what is good for them rather than what is bad for them? Put simply, the answer must be that the stimuli that they treat as "rewards" tend to be associated with things that are good for them, while the stimuli that they treat as "punishments" tend to be associated with things that are bad for them. A reward is defined as a stimulus which, when encountered, tends to lead to a repetition of the preceding behaviour pattern or patterns. A wheat grain is a rewarding stimulus in that a chick, having eaten a wheat grain, thereupon becomes more likely to eat another one if given the chance. Or if the chick is given a wheat grain every time it pecks a blue key, and this causes it to learn to peck the blue key, we say that wheat grains are rewarding. Conversely if an animal learns to *avoid* pecking the key that causes it to be given an electric shock, we say that the electric shock is punishing.

We tend to take it for granted that animals treat as rewards those things, like wheat grains, that are genuinely nutritious or otherwise good for them. And we think it almost obvious that pain will tend to signify things that are genuinely bad for the animal. But actually it is not obvious. It is easy to imagine an animal whose nervous system was wired up so that it enjoyed being injured. Something had to see to it that nervous systems treat good things as rewards and bad things as punishments, and that something was natural selection. Any genetic tendency to enjoy injury will be propagated less successfully than the opposite genetic tendency. If wheat is the only available staple diet for the species, any individual with a genetic tendency to treat the taste of wheat as punishment would not survive long. In order for the changes that we call learning to be changes in the right direction, the animal must be provided with the right list of reward and punishment stimuli, and this provisioning is ultimately accomplished by natural selection. This is what Konrad Lorenz meant when he referred to innate teaching mechanisms or the "innate schoolmarm".

If, then, the wisdom that an animal acquires during a lifetime of experience could somehow be written into its genes and bequeathed to its offspring, those offspring might indeed benefit. But such a process could not, in itself, be the basis for progressive evolution in adaptive directions. The progress and direction all have to come from natural selection, even if selection acts indirectly via mechanisms of reward and punishment. Lamarck was right to see his theory as incomplete without an additional guiding force. The guiding force that he suggested—"needs" automatically calling forth the means to their satisfaction—is obviously mystical nonsense: the theory simply assumes what it sets out to explain, and is therefore no better than God. But Lamarck deserves credit for realising that *some* guiding force was necessary over and above mere use and disuse. I suspect that if he had lived longer he would have welcomed natural selection as a guiding force that could actually work.

So much, then, for my first reason for doubting that a form of life could exist based on Lamarckian evolution. My second doubt concerns the capacity of use and disuse to put together complex functioning organs. It is one thing to agree that the size of an organ, or of a part of an organ, could be influenced by how much it is used. Maybe an animal that uses its arms a lot will get big arm muscles, while one that uses its legs will get big leg muscles. But here we are talking about a very simple change in size. Can the same principle work for subtle improvements in complex organs like eyes, ears or hearts? If an animal uses its eyes a great deal, if the total number of photons passing through the lens increases, can this improve the optical qualities of the lens? Can mere exercise promote the correction of spherical and chromatic aberration? Certainly not.

Tony Cooke/Oxford Scientific Films

The ostrich has calluses before it is born, but Darwinian adaptation can account for this perfectly well

The theory of natural selection has no trouble in explaining the evolution of adaptive improvements of indefinite complexity. A lens with spherical aberration may well be significantly more likely to lead to the animal's death than a lens with slightly less spherical aberration: a slight misjudgement of the exact position of a cliff could easily be fatal to a fast-flying animal such as a swift. Any change whatsoever, however small, subtle and indirect, that increases the ability of a swift to avoid collisions, any slight improvement in its eyes or its brain or anywhere in its body, will be favoured by natural selection. In the theory of natural selection, the relationship between adaptation and evolution is a direct and close-coupled one: anything that is good for the animal will tend to evolve. In the Lamarckian theory, the relationship is tenuous and cloudy, diffused through a crude smokescreen of "use and disuse". It relies on the rule: the more the animal uses a certain bit of itself, the bigger that bit ought to be. That rule just might have some sort of minimal validity in some circumstances. But it will lack general applicability, and, in any case, as a sculptor of adaptation it is a crude bludgeon in comparison to the fine sharp chisels of natural selection.

If life is ever discovered on another planet, therefore, I believe we can be pretty sure that it will have evolved by some form of Darwinian mechanism. The directional element in the evolution, the thrust in the direction of adaptedness, will have been produced by non-random survival from an initially random (or at least undirected) pool of variation. If the inheritance of acquired characteristics does take place on some planet, it will serve, at best, to smooth the path of adaptive change between generations: it cannot introduce new adaptive complexity.

PART VII
THE MAN AND HIS IDEAS

Darwin's ideas about evolution and natural selection changed just as surely as species change. We trace the evolution of natural selection and offer a portrait of Darwin written in 1958 to commemorate the centenary of the presentation of natural selection to the Linnean Society. Finally, to mark the current centenary, a look at the death of Darwin

Chapter Twenty

The path to natural selection

Darwin's theory, like the facts he sought to explain, shows evidence of change

MICHAEL GHISELIN

HOW and when did Charles Darwin discover natural selection? The topic remains full of puzzles, but it is now possible to give a fairly accurate and reasonably detailed account of what happened. This replaces a lot of mythology and misconception, much of which goes back to Darwin himself. According to the traditional account, Darwin's evolutionary researches began during the voyage of the Beagle. Upon his return to England he began an "inductive" search for a theory, patiently accumulating facts and studying the modification of domesticated animals and plants. Then, in late September of 1838, he just happened to read a book by Thomas Malthus, and all of a sudden he had a theory with which to work.

This account is not entirely wrong, but it is oversimplified, misleading, and grossly inaccurate in many respects. We now appreciate the fact that Darwin was basically a theoretician and a synthesiser. He was a dedicated empiricist, but when he dealt with facts, there were always theoretical issues in mind. He may have spent years circumnavigating the globe and more years sequestered at Down, but he was in no sense intellectually isolated. He knew the literature, both old and current, on a vast range of subjects and was, indeed, one of the most impressive polymaths of all time. The sheer volume of notes and correspondence makes one wonder when the man slept—but supposedly he lost years of work to ill health. We should not blame Darwin for the mythology, however. He was simply engaging in the conventional dissimulation of his time and telling people what they wanted to hear. Pretending that his discovery had emerged from the facts made it appear more plausible to his contemporaries. In those days one was supposed to be a "Baconian", a follower of Francis Bacon, just as now one claims to follow Karl Popper. It was the sort of academic salesmanship that will always be with us.

Chris Corr

Down House, Kent, where Darwin lived and worked

We are fortunate in having a rich supply of documents that enables us to piece together what really happened. These include the published works and a great deal of manuscript and other material at Cambridge University, material that is accessible largely thanks to the hard work and inspired leadership of Sydney Smith. My account is fairly close to a consensus of recent study on the main points, and draws heavily on the contributions of Sandra Herbert, M. J. S. Hodge, David Kohn, Dov Ospovat, Silvan S. Schweber and Frank M. Sulloway. The novelty consists mostly of some observations on Darwin's psychological investigations that help to complete the picture. While Darwin was discovering natural selection, he was also laying the foundation for a new approach to psychology. The two endeavours were closely interconnected and lent each other mutual support. Darwin is widely recognised as a comparative and experimental psychologist second to none. It is not so well appreciated how far back this interest went or how pervasive was its influence on Darwin's work. The Beagle research included a great deal of work on animal behaviour, but he was already taking notes on such matters while still a medical student at Edinburgh. In *The Expression of the Emotions* he tells us of betting a group of young men that they could not sneeze by taking snuff, and winning. Probably, these were the brilliant scholars with whom he kept company at Cambridge as a student. (Sydney Smith has found that they were not the lazy fellows depicted in legend.)

At the beginning of his voyage on the Beagle (27 December, 1831, to 2 October, 1836), Darwin planned to devote most of his time to geology and marine invertebrate zoology. Influenced by Charles Lyell's *Principles of Geology* and encouraged by the success of his work in that area, Darwin came to emphasise geology in his research. This had important effects on his biology, not the least of which was providing experience in thinking in historical terms. As Hodge points out, much of Darwin's biology can be viewed as an effort to develop Lyell's approach to biogeography, which was quite novel. In South America, Darwin found compelling evidence that the land and climate had changed, and so had the biota. Gigantic land mammals had become extinct, but shells had scarcely changed if at all. Introduced species had crowded out the old ones. According to Lyell, change in fauna and flora could largely be explained in terms of organisms having moved from place to place, limited by barriers and ecological influences. The aim of biogeography, as Darwin was developing it, was to trace species to their places of origin in the light of their capacity for dispersal and the history of geological change.

Darwin was thus provided with a biology that was historical, but not evolutionary. Species would still have to be created somewhere. Two additional elements were yet to be added. The first was the possibility of unlimited change in character—transmutation. The second was the derivation of new species from old ones through fragmentation—what we now call speciation. A limited amount of transmutation was already recognised: the formation of local races. Darwin studied varieties on a geographic basis and evidently took it for granted that local conditions somehow modified the organisms—by analogy with physiological adaptation. Already, however, he was concerned with another important mechanism of change—behavioural adaptation, especially habits becoming hereditary. In the Falkland Islands he was much impressed with the tame foxes and with geese that did not fly. However, in order to do more than populate the world with lineages that adapted only locally, and to a minor degree, Darwin had to relate the taxonomic pattern to the biogeographical one. He had to realise that the data call for an extrapolation from the diversity within species to the diversity within genera and families.

Legend has it that this insight came to him at the Galapagos Islands. We now know that Darwin was not attending closely to higher-category taxonomy at that time. This became important only later, but how much later is highly debatable. It could easily have occurred towards the end of the voyage when Darwin was sorting out his notes. Perhaps it happened later. Nonetheless, Darwin was surprised to find that the animals of the Galapagos had a strikingly South American character. He attributed the local diversity to migration and the formation of local races. Another striking feature of the Galapagos was behaviour like that he had seen at the Falkland Islands. The birds were largely unafraid of man, and the marine iguana could not be driven out to sea, where its only natural enemies occurred.

Later in the voyage Darwin would visit more islands. He would see New Zealand and Australia with their peculiar faunas. He would repeatedly see local animals and plants losing out in competition with introduced species. Probably, he began to take evolution seriously around the time the Beagle visited St Helena. Upon his return to England he moved fast, especially so as to advance his reputation as a geologist. Also he put his crucial biological specimens into the hands of specialists with great despatch. The fossil mammals went to Richard Owen, the birds to John Gould, and the remaining vertebrates, the plants, insects and spiders, to others. Fossil and living shells, which related to his geological work, he gave to others—but not those marine invertebrates that had no such significance. These he reserved for himself. As he sorted out his specimens and worked on his *Journal of Researches,* and finally in receiving Gould's opinion that the Galapagos birds were species, not varieties, Darwin had more and more of a picture that cried out for evolutionary connections within genera and families.

At no point, however, was Darwin "converted" to evolution in the sense that one joins a church. His commitment developed more or less gradually, in a number of small steps at most. Nonetheless, he did consider evolution seriously enough to

Bruce Coleman/CUP

Darwin planned to devote most of his time on HMS Beagle (above) to geology. Two Galapagos tortoises (right), which helped persuade the young Darwin that species were not fixed

begin working on it in earnest around March of 1837. In July of that year he began Notebook *B*, one of a series. These notebooks served primarily as a basis for intended publications. Notebook *A* deals with geology, Notebooks *B* and *C* with all aspects of evolution—including the psychological ones. When *C* was finished, Darwin split the series. *M* is concerned with evolutionary psychology and related topics. Why *M*? That *M* stands for mind, metaphysics or materialism is a popular conjecture. I suspect, however, that Darwin merely went to the middle of the alphabet. Notebooks *D* and *E* treat evolutionary biology in general. The crucial reading of Malthus occurred towards the end of September 1838, when Notebook *D* was almost full. Notebooks *E* and *N* are both post-Malthusian and begin in early October of 1838.

Notebook *B* begins with what Hodge calls a "zoonomical sketch"—an outline for a theory on the "laws of life". This can only mean that Darwin was committed enough to evolution that he hoped to publish something on it. He already had a theory and referred to it as such, long before he had thought of natural selection. This theory has much in common with those of Darwin's well-known predecessors, Erasmus Darwin and Jean Baptiste de Lamarck. It was not, strictly speaking, replaced by natural selection; it was in part replaced, in part modified, and in part incorporated into Darwin's later versions of evolutionary biology. As he worked on this theory, he developed other views that went far beyond just a mechanism for evolution.

The first theory was based upon an analogy between organisms and species. He thought that both have a life span, and that this helped to explain extinction. Darwin had been considering this possibility for quite some time, but it is hard to say just how long. Sexual reproduction, he thought, produced variants that tended to be adaptive; whereas asexual reproduction produced only replicates of the parent. Thus sex generated variability more or less as some of us now interpret it—but without the selection that makes it seem a plausible mechanism for adaptation. We can trace Darwin's enormous contribution to the biology of reproduction—he wrote several books on that topic—to this hypothesis. Darwin at that time had another explanation for adaptive change. It resulted from change in behaviour, and this was something Darwin never abandoned. He thought, for example, that a species could be formed as a result of "repugnance to intermarriage". Later in the series of notebooks he developed the theory that change in habit often precedes change in structure. He continued to believe that this is so, and so do we moderns. Views he developed in Notebooks *M* and *N* relate to this and are the foundations for Darwin's major contributions many years later to sexual selection, emotional expression, the evolution of social behaviour and other topics, many of which retain their interest to this day.

Darwin claimed to have developed his theory of natural selection by studying artificial selection and seeing how plant and animal breeders modified domesticated organisms through deliberate choice. Negative evidence—the paucity of references to artificial selection in the notebooks—has led to what probably is an overreaction against this assertion. At first Darwin did not consider artificial selection to be an effective mechanism for evolutionary change under natural conditions. He called it "picking" and envisioned it as a matter of pairing deviant individuals which were then kept isolated. He did, however, study the work of breeders, albeit not as a model for evolution by selection. That came later after he had discovered natural selection. He was investigating such topics as the laws of heredity, the effects of hybridisation and mating behaviour. An improved understanding of artificial selection did result, but only as a serendipitous by-product.

The main thrust of Notebooks *C* and *D* was an effort to develop a new approach to systematics based upon evolution. He looked upon himself as the great reformer of biogeography, comparative anatomy and formal taxonomy. His readings indicated that his theories about evolutionary mechanisms were not all that new. This effort was hardly wasted. After all, it was completely successful, even though evolutionary classification is still resisted in some quarters (see chapter ten). His views on all aspects of evolutionary biology, nonetheless, continued to evolve. He became more critical of teleology. His conception of the natural economy increasingly emphasised competition. His study of the "wars of organic being" predisposed him to appreciate something very different—the struggle for existence, or what we call reproductive competition.

Darwin read Malthus late in September of 1838—not for "amusement" as he put it in his so-called autobiography, but as an integral part of his work on the evolution of morals, work that is the direct ancestor of modern sociobiology. As Silvan Schweber has emphasised, Darwin in this period of his life was reading about all sorts of things, including philosophy and political economy. The vast scope of his erudition contributed as much to his success as did his detailed knowledge of the facts. It has long been a puzzle just what Malthus did for Darwin. In retrospect it is now fairly easy to see why there is a puzzle, and why it continues to engage people.

Darwin invented the modern science of ecology. G. F. Gause and Vito Volterra, the founders of mathematical population biology, freely acknowledge their debt to him. Anyone equipped with a sound knowledge of late 20th-century population biology can go back to Darwin's writing on such matters and find them readily intelligible, even though we now use a rather different jargon. However, the fundamental principles of evolutionary ecology were long forgotten, and their rediscovery has taken place only in the past 20 years or so. Those principles are not now, and have never been, common knowledge. It stands to reason that philosophers, historians and biologists without the requisite expertise would find Darwin's views unintelligible. They would naturally think of such words as "struggle" in anthropomorphic terms. Thus when Darwin explains how populations respond to density-dependent selection, ecologists are astounded by his prescience, and philosophers take him for a fool. Economists had a much better understanding of competition than biologists did in 1838, and to some extent this is still the case.

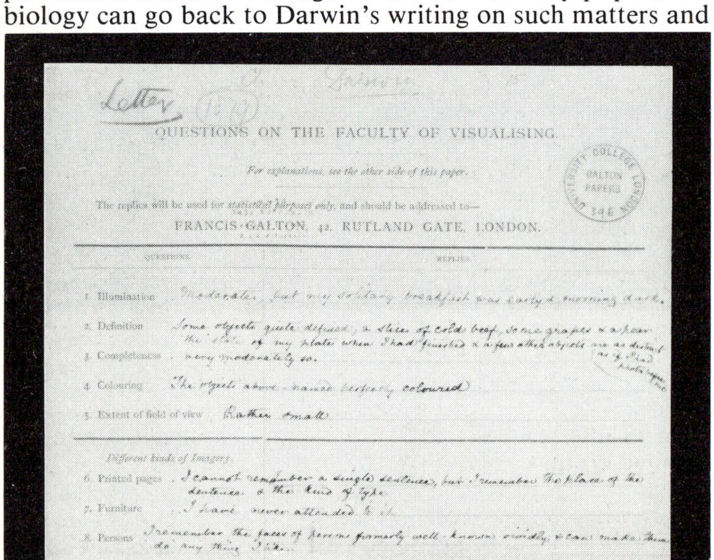
Letter

QUESTIONS ON THE FACULTY OF VISUALISING.

For explanations, see the other side of this paper.

The replies will be used for statistical purposes only, and should be addressed to—

FRANCIS GALTON, 42, RUTLAND GATE, LONDON.

GALTON PAPERS

QUESTIONS	REPLIES
1. Illumination	Moderate, but my solitary breakfast was early & morning dark.
2. Definition	Some objects quite defined, a slice of cold beef, some grapes & a pear the state of my plate when I had finished & a few other objects are as distinct as if I had photographed
3. Completeness	very moderately so.
4. Colouring	The objects above named perfectly coloured
5. Extent of field of view	Rather small
Different kinds of Imagery.	
6. Printed pages	I cannot remember a single sentence, but I remember the place of the sentence & the kind of type.
7. Furniture	I have never attended to it.
8. Persons	I remember the faces of persons formerly well known vividly, & can make them do any thing I like.
9. Scenery	Remembrance vivid & distinct & gives me pleasure.
10. Geography	No

Darwin's reply to a questionnaire from Francis Galton. Galton, the statistician and psychologist, was a cousin of Darwin's

Reading Malthus helped Darwin to understand the structure and processes of populations. Previously, he was thinking about how one group of organisms might make another extinct. There would be some sort of differential survival in response to the conditions of existence. This, however, is not natural selection. It does not explain why one group comes to be more successful than another. Population pressure gives the answer. The environment tends to become saturated with a variety of competing organisms. Under such conditions, any organism that can exploit an under-utilised portion of the "natural economy" will out-reproduce other organisms of its species. In other words, competition forces individuals into empty niches, dragging their species with them. Any variant, however small, that had this effect would tend to accumulate through successive generations. A mere sorting out of species would have no such effect—at best there could be a crude fit between the properties of the group and the prevailing average conditions of existence. This provided the clue to the right way to think about selection: as something that goes on between individuals, not species.

A lot of changes were in order, but these accumulated gradually over a period of months and even years. Darwin left some early traces of his reorganised investigation in Notebooks *E* and *N*. He attributed a new role to sex. Interbreeding allows sex to affect each species as a whole, so that evolution is not just the outcome of successive change in single lineages. He thought this would produce a closer adaptive fit of all aspects of the organisms' biology to the environment. He realised that artificial selection works in basically the same way that natural selection does—many small differences are chosen and incorporated within the breed. This made the analogy a valuable way of solving problems. It dawned upon him that combat between males and the attraction of peahens for peacocks' tails would have an effect like that of artificial selection—and so he discovered sexual selection. Nor did it take long for him to realise that instincts could evolve, not just by inherited habits but by natural selection of fortuitous hereditary variations in the nervous system.

The evidence clearly shows that Darwin's theory evolved. We need not posit some analogue of special creation to account for its origin—and no occult faculties such as inspiration. There was no "orthogenesis"—no straight-line analogue of a developing embryo. There was no "catastrophism"—no sudden replacement of one theory by a totally different one, and the two supposedly incommensurable. There were no "saltations"—no great leaps of the intellect. There were indeed periods of stasis and periods of rapid change—the "punctuated equilibrium" hypothesis of Niles Eldredge and Stephen Jay Gould fits quite well. And views were modified to such an extent that without a record of the intermediate stages we would be hard-pressed to see the continuity.

The history of science is best intelligible as closely analogous to organic evolution. The documents are like the fossil record: non-evolutionary models for the history of science result from an inadequate appreciation of the gaps. For Darwin we are lucky that such a wealth of documents has been preserved. It looks as if his kind of biology will continue to do far more than just account for the origin of species.

Chapter Twenty-one

Darwin's greatest work

Charles Darwin was 'an outstanding example of the fact that in education and training the race is not always to the swift'. In *The Origin of Species* he produced a book that revolutionised human thought

EDWARD HINDLE

MEN of science celebrated the centenary of a communication made to the Linnean Society of London at its evening meeting on 1 July, 1858. This society holds its meetings in Burlington House in a room to the left of the entrance archway, and on this occasion a joint paper entitled "On the tendency of species to form varieties; and on the perpetuation of varieties and species by natural means of selection" was communicated by Messrs C. Darwin and A. Wallace. There was no discussion and the only published comment on the meeting was by Professor James Haughton, of Dublin, whose verdict was that all that was new was false and what was true was old.

Certainly none of those present could have had any idea of the tremendous consequences that would result from the joint paper. Within a few years it had revolutionised human thought, and it must be regarded as one of the most important communications ever made to a learned society. Neither of the authors could be considered in any way revolutionary: one a well-to-do country gentleman of cultured tastes, with all the advantages of a university education, known to a select few as a first-class naturalist from his account of the natural history and geology of the countries visited during his voyage round the world in HMS Beagle; the other of very different background, also well travelled, but gaining a precarious livelihood by voyaging to remote parts of the world to obtain rare insects and other animals for sale to museums and private collectors. Both had the inestimable advantage of possessing experience of the struggle for life in teeming tropical jungles, and independently both had arrived at the same theory of evolution by natural selection, though by rather different routes—Charles Darwin as the result of patient and laborious accumulation of facts extending over more than 20 years, Wallace by a flash of inspiration.

Four different Galapagos finches, four different beaks

Darwin might almost be said to have inherited an interest in evolution, for his grandfather, Erasmus Darwin, in his book *Zoonomia* (1794), expounded views similar to those of Johann Goethe and Jean Baptiste de Lamarck that species were not immutable. Already in 1837 Charles had written "Transmutation of species" at the head of a notebook and begun to collect facts concerning species and varieties, especially the results of breeding experiments. The following year he read Thomas Malthus's *Essay on Population* dealing with the struggle for existence among human populations, and it at once struck him that a similar struggle among animals and plants might result in favourable variations tending to be preserved and unfavourable ones to be destroyed. His observations during the voyage of the Beagle had already had a profound influence on the growth of his evolutionary ideas, in particular his visit to the "eminently curious" Galapagos.

This archipelago is zoologically a part of South America, but half the plants and many of the animals are so different that they had to be classified as new species. The finches are of especial interest, for although obviously closely related they have become separated off into at least 13 distinct species showing extraordinary modifications of the beak in accordance with different modes of life. In one species the beak has become large and strong like that of a hawfinch, and at the other extreme there is another species with the slender bill of a warbler. All these species are peculiar to the Galapagos and the separation of these islands from the mainland was the only light Darwin could throw on the production of so many new species. The full details concerning the differences in the birds of these small, barren and rocky islands and the way they had grown apart were entered in his notebook followed by the comment that such facts "would undermine the stability of species".

Darwin continued to gather material when he returned to England, and in 1842 wrote a brief extract of his theory. At that time orthodox opinion held that every species of animal and plant, both living and extinct, had been separately created and that the animals at least were descendants of pairs which had survived the deluge by entering Noah's Ark. To hold a contrary opinion was, to use his own words, "like confessing a murder", but early in 1844 Darwin took his friend Sir Joseph Dalton Hooker into his confidence and that summer enlarged his brief extract into a longer essay in two parts—the first, "On the variation of organic beings under domestication and in their natural state"; and the second, "On the evidence favourable and opposed to the view that species are naturally formed races descended from common stocks". A copy of this essay was sent to Hooker, and then for many years Darwin devoted himself to working out the results of the voyage of the Beagle, also to geological observations and the preparation of his great monograph on barnacles.

The species problem had never been forgotten, however, and during the years he collected all the facts he could find concerning variation, the results of hybridisation, geological succession of fossils, geographical distribution, classification, comparative anatomy, embryology, vestigial organs and similar topics. Finally, in 1856, he started the preparation of a book in which his ideas could be fully presented together with supporting evidence. Both Sir Joseph Hooker and Sir Charles Lyell urged Darwin to hurry with this book, but there was always some new problem to be solved and by 1858 he had finished only 11 chapters. Then on 18 June of that year an essay arrived from Alfred Russel Wallace, together with a letter in which he asked Darwin to give his opinion of the essay, and if he thought well of it to send it on to Lyell. This essay, entitled "On the tendency of varieties to depart indefinitely from the original type", showed that by a remarkable coincidence Wallace had independently and on similar grounds arrived at the same conclusions concerning evolution.

Darwin at once wrote to Charles Lyell: "Your words have come true with a vengeance—that I should be forestalled—I never saw a more striking coincidence . . . So all my originality, whatever it may amount to, will be smashed, though my book, if it will ever have any value, will not be deteriorated, as all the labour consists in the application of the theory." Darwin so highly appreciated Wallace's essay

National Portrait Gallery

Alfred Russel Wallace

that he proposed to have it published at once in one of the scientific journals, but Hooker and Lyell decided that in the interests of science a joint publication of both Darwin's and Wallace's work was essential. Accordingly it was arranged that relevant portions of both essays should be read at a meeting of the Linnean Society.

There was never any trace of rivalry between these two great men and Darwin wrote to Lyell: "I would far rather burn my whole book than that he [Wallace] or any other man should think that I had behaved in a paltry spirit." Noble words, but Darwin admitted in a letter to Hooker: "I have always thought it very possible that I might be forestalled, but I fancied that I had a grand enough soul not to care; but I found myself mistaken and punished. I had, however, resigned myself and had half written a letter to Wallace to give up all priority to him and should certainly not have changed it if it had not been for Lyell's and your quite extraordinary kindness. I am more than satisfied at what took place at the Linnean Society."

In September 1858 Darwin set to work to prepare a volume on the transmutation of species, based on the MS he had begun on a much larger scale in 1856. It cost 13 months and 10 days hard labour and was published under the title of *The Origin of Species* in November 1859. Darwin himself attributed the success of this book to having long before written two condensed sketches, and having finally abstracted a much larger manuscript. Another element was its moderate size which was credited to the appearance of Wallace's essay, for his original intention had been to write a book four or five times as large as the *Origin*, which very few would have had the patience to read.

In the first complete edition of his autobiography, edited by his granddaughter Nora Barlow, Darwin states: "I gained much by my delay in publishing from about 1839 when the theory was clearly conceived, to 1859; and I lost nothing by it, for I cared very little whether men attributed most originality to me or Wallace; and his essay no doubt aided in the reception of the theory."

This was undoubtedly Darwin's greatest work, as it provided the principles on which it can be understood how living beings, including man, came to be what they are, and completely altered the course of human thought. It continues to throw light into the most diverse fields—art, literature, music, law, science and technology, history, sociology—and recently the concept of evolution has been applied to the subject of ethics.

Even if he had written nothing else, this book would have been enough to stamp Darwin as a genius, but he continued throughout his life to produce major works all distinguished by a profound insight into natural problems. His other main publications include the *Naturalist's Voyage in HMS Beagle* (1839), a *Monograph on the Cirrepedia (Barnacles)* (1851-54), *The Structure and Distribution of Coral Reefs* (1842), *The Fertilisation of Orchids* (1862), *On the Two Forms or Dimorphic Condition of Primula* (1862), *Climbing Plants* (1864), *Variation of Animals and Plants under Domestication* (1868), *Descent of Man* (1871), *Expression of the Emotions in Man and Animals* (1872), *Insectivorous Plants* (1875), *Effects of Cross- and Self-Fertilisation in the Vegetable Kingdom* (1876), *The Different Forms of Flowers* (1877), *Power of Movement in Plants* (1880), and *The Formation of Vegetable Mould through the Action of Worms* (1881).

Darwin is an interesting example of late development, for it is doubtful whether under modern conditions he could have entered any British university. He considered his time at Cambridge was wasted as far as academical studies were concerned. However, his time was far from wasted, for a growing interest in scientific work was encouraged by his friendship with John Henslow, professor of botany at Cambridge. He also began the study of geology, and collected beetles, but Alexander von Humboldt and Sir John Herschel inspired him with a burning zeal to add even the most humble contribution to Natural Science. During the voyage of the Beagle he was able to indulge his youthful pleasure in direct observation and collecting, and this developed into the more mature satisfaction of the theorist.

The career of this simple, kindly and very great man, who devoted his life with such success to the pursuit of truth, is an outstanding example of the fact that in education and training the race is not always to the swift.

Written in 1958 to commemorate the centenary of the presentation to the Linnean Society.

Chapter Twenty-two

The death of Darwin

From establishment scourge to establishment member, Charles Darwin's path from *The Origin* to his resting place next to Isaac Newton in Westminster Abbey

IRVING STONE

SOME are forgotten the moment the first spadeful of earth hits the coffin. Charles Darwin will live as long as there are men on Earth to think, analyse, read, write and observe.

Why? Because Darwin, along with Albert Einstein, Karl Marx and Sigmund Freud, created our modern world. Though he is likely to remain a controversial figure—those whose lives are centred in religion will continue to view him as an enemy of their God—he has become acknowledged since his death as the foremost scientific philosopher, whose findings opened the heavy, locked doors of all species on Earth, past and present; of man's origin and descent.

As far as time span is concerned, Darwin lived a normal life; he was born on 12 February, 1809, and died on 19 April, 1882. At the age of 73 his death was neither unexpected nor tragic. His almost incredible body of work had already been accomplished and set forth in print.

Unlike his outstanding associates and closest friends, the botanist Joseph Hooker and Charles Lyell the geologist, he was never knighted. The government of England preferred to avoid the outcry from those who believed that the six days of *Genesis* accurately and adequately described the creation of the Universe. Yet all official opposition of Church and State was stilled for the momentous occasion of his funeral. In what amounted to a revelation, the Dean of Westminster chose for Darwin's burial spot in the abbey a place on the north side of the nave, close to the angle of the choir screen, near the grave of Sir Isaac Newton. The funeral was one of the most colourful in English history.

Yet Emma Wedgwood, Darwin's wife for 43 years, had wanted a simple funeral in the ancient graveyard adjoining the little church at Downe, where Charles would lie near his brother Erasmus, his children Mary Eleanor and Charles Waring and where Emma herself would one day join them.

The story of how and why Charles Darwin was buried with pomp and ceremony in Westminster Abbey is a stirring one. But first it may be interesting to take a look at Darwin's health, and how he had *not* died a number of times, considerably earlier, as he fondly imagined he would. (I use the word "fondly" advisedly.)

Hulton Picture Library

The last photograph

Charles Darwin grew to a shade under six feet, a lean, sinewy, powerful and inexhaustible young man, walking miles every day along the River Severn. He went geologising in North Wales with Adam Sedgwick, and spent robust weekends collecting in the fens of Gamlingay with his mentor, John Henslow. Sedgwick and Henslow were both professors at Cambridge, where Darwin was an undergraduate at Christ's College. On the five-year journey of HMS Beagle, he was an indefatigable non-member of the crew (he was signed on as an "unfinished naturalist"). He trudged all one night to rescue his dehydrated friend and "beau ideal of a captain", Robert Fitzroy, and accomplished what few had ever dared before him, traversing the Andes on foot and on horseback, crossing the pampas of Patagonia.

Darwin returned to England at 27 in a robust state of mind and body. It was not until a year later, when he began to write in his evolutionary notebooks, that he first felt and commented on his illnesses, forcing himself into a lifetime of severe, repugnant and sometimes ludicrous disability. He survived them all . . . barely. The miracle is that he did not kill himself with his cures.

Is there any evidence that Charles Darwin was ever physically ill? Dr Ralph Cox Jr, in his exhaustive book, *To be an Invalid,* lists six most widely discussed possibilities of cause for Darwin's ailments. His use of arsenic for skin outbreaks on his fingers and lips during his collecting in the fens of Lincolnshire, while a student at Cambridge. The heart palpitations in Plymouth, during the trying months of bad weather before HMS Beagle could set sail. A possible injury to his digestive tract during his years of devastating seasickness aboard Beagle. An alleged heat stroke on 2 October, 1833, while riding horseback from Buenos Aires to Santa Fe in Argentina. The drinking of raw chichi wine (whisky) in September 1834 while visiting a gold mine in Chile, which he said "half-poisoned him"; he was very ill for a month but was cured by rest and the calomel prescribed by his friend Dr Benjamin Bynoe, surgeon on the Beagle. The effect of Chagas' disease from a bite by the "great black bug of the Pampas" (*Triatoma infestans*) on 25 March, 1835, while in the village of Luxan in Argentina. Colp then demolishes each argument.

In short, there was nothing organically or medically wrong with Charles Darwin, excepting that during his 45 years of search and discovery he knew for sure what another student at Christ's College, John Milton, had declared 200 years before, "All Hell broke loose".

Martyrs over the centuries have been burned at the stake, their agony and death embracing perhaps an hour. Charles Darwin was drawn and quartered, disembowelled, eaten while pilloried to a live ant hill from the moment of publication of *The Origin of Species* in 1859, a seemingly unending period of 23 years. He appears to have been the single most abused individual in all of England's stormy history: a proclaimed beast, a pervert, an anti-Christ, a moral leper in public sermon, lecture, newspaper and magazine article, in books, decrees, promulgations, few of which he was allowed to ignore or escape.

The wonder is that he survived the calumny until his 73rd year. And it makes understandable some of the cures he undertook: a bland diet with lots of wine, abstention from snuff, wrapping himself in alternating brass and zinc wires moistened with vinegar, hydroelectric chains, patent medicines that blatantly touted their curative qualities in the London newspapers. One of the fashionable cures of the day—the hydropathy spas of Dr James Gully at Malvern and Dr Edward Lane at Moor Park—actually helped him. When he lay wrapped in cold wet sheets for hours, or submerged in cold baths, he could not only not work, but also not think about the demon of evolution he had released upon an unwilling and unprepared world. At his 60th birthday party Dr Henry Holland, his cousin and satiric critic, and also family physician, commented: "Whoever would have thought this possible? You've been expecting an imminent death for several decades!"

All of this had happened to an easygoing, friendly, happy-go-lucky lad without any ambition except to collect beetles and, in September of each year, shoot partridge at his Uncle Jos Wedgwood's estate, Maer Hall. When he described himself in his letters as an invalid, his friends and enemies alike called him a hypochondriac. Even his granddaughter Lady Nora Barlow, who edited Charles Darwin's autobiography, told me that her grandfather was a hypochondriac and that his descendants were as well.

They all were wrong. Charles Darwin suffered psychosomatically induced illnesses, including the green bile that he vomited, between meals, for years on end. What he was actually rejecting was the bile of abuse and calumny relentlessly poured upon his peace-loving head. Nature had simply chosen the wrong disposition for the task; but ultimately the right man, the one who could get the whole job of the evolution and natural selection of species on so solid and provable a base that decades of determined opposition could not overthrow its working conclusions, and naturalists of the future could continue the revealing and rewarding work.

Darwin noticed his serious heart palpitations and weak pulse in 1882, after the success of his last book, *The Formation of Vegetable Mould Through the Action of Worms.* When the life and times of earthworms had become the subject of the day, he said: "My book has been received with almost laughable enthusiasm". It was Charles Darwin's last enthusiasm. His London physician, Dr Clark, told him that "there is some derangement of the heart, but it is not serious".

Darwin knew better. He had a premonition that his end was near. For the first time, he was right. A few days later he had a seizure while walking to the home of a friend in London. Emma put him to bed, gave him quinine, which helped him, so that he was able to write two short papers for the Linnean Society about his studies on roots and chlorophyll bodies. Then he began slipping away. His last words to Emma were: "I'm not afraid to die".

Upon word of his death, his detractors circulated a rumour that he had repented on his deathbed, and asked God's forgiveness for his blasphemies. There was not an iota of truth to the charge, yet it still surfaces today, presented as fact by those who would like to believe it.

The family plan to bury Charles Darwin quietly at Downe was thwarted by Darwin's long-time friend, neighbour and student, John Lubbock, banker and naturalist, and a member of parliament. He persuaded Emma that "from a national point of view, it is clearly right that he should be buried in the abbey". He also induced 20 members of parliament to sign a petition to Dr George Bradley, Dean of Westminster.

HOUSE OF COMMONS, 21 APRIL, 1882

Very Rev. Sir,—We hope you will not think we are taking a liberty if we venture to suggest that it would be acceptable to a very large number of our fellow-countrymen of all classes and opinions that our illustrious countryman, Dr Darwin, should be buried in Westminster Abbey.

The funeral in Westminster Abbey took place on 26 April, 1882. Though all were warned to wear mourning clothes, there was little sadness in the abbey. Rather, there was rejoicing that Charles Darwin had lived and worked and torn veils of illusion from the misty eyes of his fellow men.

Charles Darwin had the pallbearers he would have wanted. Joseph Hooker, his dearest friend; Thomas Huxley, "Darwin's Bulldog"; John Lubbock; Alfred Wallace, who had reaffirmed that "the origin of species belongs solely to Mr Darwin". In the procession the Crown came first, then the mourner canon, the canons, the Dean's verger, the family, their servants; then the scientific bodies, representatives of the United States, Russia, all the countries of Europe; the universities, the Royal Society, friends and admirers.

It was a ceremony that would have pleased Charles Darwin. What would not have pleased him was that the following day a company of clockmakers in Fleet Street, Barraud and Lunds, closed its long-time account with Martin's Bank because R. B. Martin had attended the funeral, thereby giving the impression of endorsing Darwin's heretical theories. The last lines of the letter read: "The Day is hurrying on, which shall burn as a Fire. We shall see then who has spoken Truth, God or Mr Darwin".

Looking back from a vantage point 123 years later, one can imagine the intensity of the shock of *The Origin of Species* first on the British people, then on the Europeans and Americans. The publication of *The Origin of Species* in 1859 appeared to constitute a genuine threat to the establishment, for it became known that Charles Darwin did not believe in an anthropomorphic god, the personal, speaking God of *Genesis* who created the Universe whole in six days. Darwin had observed the creative as well as the destructive forces at work in nature, from the movement of seas and mountains, the fish fossils found high in the mountains, Earth-walking megatheriums embedded in cliffs, to the fertilisation of orchids, unknown before his investigation.

Only Darwin's fellow naturalists, some of them working in the remotest areas of the Far East, had any clue to the revolution that was about to take place in the Western world. A large part of what the school and family had been teaching about the beginning of their world, and the relationship of species, was now declared to be false. Almost the whole population was traumatised. How could one live without the assurance of a personal God fully responsible for each and everyone; who kept a beneficent watch over one's daily life and answered one's prayers? How could one accept the fact that the Universe had not been created in 4004 BC, when Darwin was insisting that all life on Earth was millions of years old, and had descended through an incredible series of mutations? A major part of the human mind had to be made over. Whole new thought patterns had to be superimposed upon old and discarded ones. It would take an heroic effort to accomplish such a feat.

Yet the shock of the new, even when it seems heretical and at first blush insane, can slowly be absorbed into the human mind, albeit painfully. Day by day and year by year, during the 23 years between the publication of *The Origin of Species* and Darwin's death, teachers were re-educated, textbooks rewritten, new courses of science introduced into schools and universities, newspaper and journal editors persuaded. As Darwin continued to publish corroborating books—*The Descent of Man* (1871), *The Effects of Cross- and Self-Fertilisation in the Vegetable Kingdom* (1876)—he acquired more and more readers, more and more believers. He was helped enormously by his co-workers, Hooker, Lyell, Thomas Huxley in England, Asa Gray in the United States, Ernst Haeckel in Germany.

Today most of the educated world accepts natural selection and the evolution of species. There is research still to be done, sorting out the details of natural selection. And there is the controversy over his major claim, that the origin of species is no more than natural selection writ large. But Darwin's theses are no longer disputed, except by the fundamentalists, who religiously insist that the Universe was created some 9000 years ago. In the United States they periodically become a virulent and destructive force, getting laws passed through their individual legislatures to put an arbitrary religious teaching in the public schools, destroying science textbooks by censorship and imperilling the separation of Church and State. For every poison there is an antidote. The Supreme Court of Arkansas has thrown out the state legislature's bill as unconstitutional (see chapter two).

The English are amused by America's scientific creationists who believe and insist that the public schools teach that the Universe and all therein was created 9000 years ago by a six-day act of God. But of course the English went through that years ago when Charles Darwin had no protection against such minds.

His theories today are the foundation stones of all the natural sciences. Few such turnabouts have been accomplished so quickly in human history. Charles Darwin was the unwilling victim of his own great brain. He had not intended to publish during his lifetime.

We honour him today, a hundred years after his death, as they honoured him on 26 April, 1882, in Westminster Abbey. No one deserves it more.

The man who had taught us that there was no such thing as an individual, separate species or genus contradicted himself in his lifework by becoming *sui generis*.

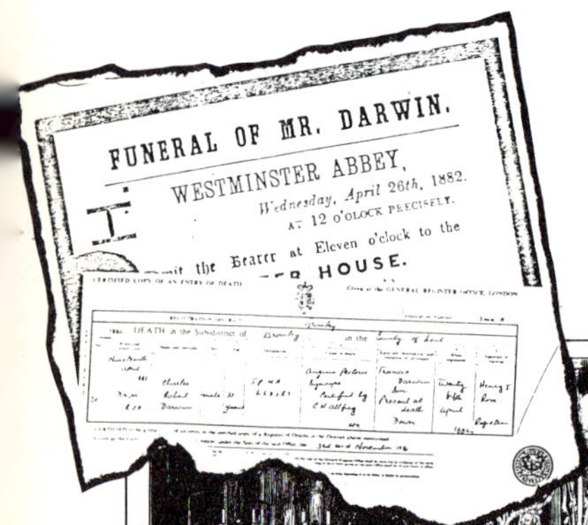

FUNERAL OF THE LATE CHARLES ROBERT DARWIN IN WESTMINSTER ABBEY

The famous and the not so famous, from the church, from science, and from governments, came to mourn Charles Robert Darwin at his funeral in Westminster Abbey. His pall-bearers were the men he loved and admired most: Joseph Hooker, Thomas Huxley, John Lubbock, and Alfred Wallace

Illustrated London News

CONTRIBUTORS

Dr Charles Carter is associate professor of biochemistry at the University of North Carolina at Chapel Hill.

Dr Brian Charlesworth is a population biologist at the School of Biological Sciences, Sussex University.

Dr Jeremy Cherfas is life sciences consultant to *New Scientist* and departmental demonstrator in animal behaviour at Oxford University.

Dr Richard Dawkins is a lecturer in the Zoology Department at Oxford University.

Dr Michael T. Ghiselin is at the Department of Biology, University of Utah.

Professor Stephen Jay Gould teaches biology, geology and history of science at Harvard University.

Edward Hindle FRS lived from 1886 to 1973, and his appointments included regius professor of zoology at Glasgow from 1935-43.

Dr Tom Kemp is curator of vertebrates at the University Museum, Oxford.

Dr H. B. D. Kettlewell was a medical practitioner and later research geneticist at the Department of Zoology, Oxford University. He died in 1979.

Dr Leslie Orgel FRS is a senior researcher at the Salk Institute in California.

Dr Geoffrey Parker is reader in zoology at Liverpool University.

Dr Colin Patterson works on fossil fishes in the British Museum (Natural History).

Mark Ridley is in the Animal Behaviour Research Group, Oxford University.

Miranda Robertson is an editor at *Nature.*

Professor Michael Ruse teaches history and philosophy at the University of Guelph, Canada.

Irving Stone is author of *The Origin: A Biographical Novel of Charles Darwin.*

Dr Chris Stringer is in the Palaeontology Department of the British Museum (Natural History).

Colin Tudge is features editor of *New Scientist.*

FURTHER READING

Peter Andrews, 1982. "Hominoid evolution", *Nature*, vol 295, pp 185-6.

J. Cronin, N. Boaz, C. Stringer and Y. Rak, 1981. "Tempo and mode in hominid evolution", *Nature*, vol 292, pp 113-22.

Richard Dawkins, 1976. *The Selfish Gene,* Oxford UP, Oxford.

E. Delson, 1981. "Paleoanthropology: Pliocene and Pleistocene human evolution", *Paleobiology*, vol 7, pp 298-305.

Clair Edwin Folsome (ed), 1979. *Life: Origin and Evolution,* W. H. Freeman, San Francisco.

Wilma George, 1982. *Darwin,* Fontana, London.

Stephen Jay Gould, 1977. *Ever Since Darwin,* Andre Deutsch, London.

Stephen Jay Gould, 1980. *The Panda's Thumb,* Norton, New York.

John Gribbin and Jeremy Cherfas, 1982. *The Monkey Puzzle,* Bodley Head, London.

Willi Hennig, 1979. *Phylogenetic Systematics* (2nd ed), University of Illinois Press, Urbana.

Richard Leakey, 1980. *The Making of Mankind,* Michael Joseph, London.

Stanley L. Miller and Leslie E. Orgel, 1974. *The Origins of Life on the Earth,* Prentice-Hall, New Jersey.

Colin Patterson, 1978. *Evolution,* Routledge & Kegan Paul, London.

Michael Ruse, 1982. *Darwinism Defended: A Guide to the Evolution Controversies,* Addison Wesley, Massachusetts and London.

Steven Stanley, 1982. *The New Evolutionary Timetable,* Basic Books, New York.

E. O. Wilson, 1975. *Sociobiology: The New Synthesis,* Belknap, Massachusetts.